Rebuilding Community in Kosovo

Rebuilding Community in Kosovo

Maurice Mitchell

This edition first published in the UK in 2003 by
The Centre for Alternative Technology Charity Ltd.
Registered Charity number: 265239
Machynlleth, Powys, SY20 9AZ, UK
Tel. 01654 705980 • Fax. 01654 702782
email: info@cat.org.uk • Web: www.cat.org.uk or www.ecobooks.co.uk

ISBN 1 90217-517-4
1 2 3 4 5 6 7 8 9 10

Mail Order copies from: Buy Green By Mail Tel. 01654 705959

Printed on paper obtained from sustainable sources in
Great Britain by Cambrian Printers, Aberystwyth (01970 627111)

Contents

Acknowledgements

Books such as this do not materialise without significant support. I would first like to thank London Metropolitan University for the research funds that made the field trips and this publication possible. The British Council, under the direction of Arjeta Emra, Centre Manager in Pristina, provided resources and administrative support for work with the University of Pristina during the second field trip. I would also like to thank Denny Lane, UNMIK administrator in Vushtrri, for his energetic support and his continuous efforts to open doors to facilitate our investigations and explorations in 2000 and 2001.

The teaching programme in Kosovo was made possible by the generosity of Development Workshop (DW) in hosting the field trips we have undertaken there. Specifically I would like to thank John Norton, Director of DW, Raymond Burton, who managed the office in Vushtrii during our first visit and provided valuable teaching support during our second trip, Tim Booth and Helen Little, who ran the DW office during our second field trip and who provided both valuable advice and support in Kosovo and material for the writing of the last chapter of this book.

Architectural colleagues on the teaching staff at the University of Pristina, Dukagjin Hasimja, Ilir Gjinolli, and Shqipe and Lulezim Nixha were tirelessly generous with their time and notably hospitable, welcoming us into their exciting and vigorous new world. We are very glad that on different visits all of them were able to make a return study trip to us here in London.

My teaching partners for the Kosovo programme, Sara Cole, who taught us all the techniques of the Situationist Derive in the first year, and Rafael Marks, who took over and developed teaching methods in the second year and helped with reading drafts, provided valuable help and encouragement.

I would like to thank all the students from Kosovo and the UK who took part in the Kosovo programme. Whether or not their work is represented, this book records their thoughts and experiences. I am also grateful for the encouragement of the young architectural student from Kosovo, who in 2000 was studying at East London University and whose name I now have no record of (his photographs are reproduced on pages 1, 8 and 12).

I have listed, overleaf, the names of London Metropolitan University architectural students who took part in the Kosovo programme, some of the Pristina University students who took part and who paid a return visit to us in London (unfortunately, I am only able to acknowledge those whose names I have a record of) and those architectural students on the Masters Course: International Vernacular Architecture of the World at Oxford Brookes University who took part in design projects based on the situation in Vushtrri.

London Metroplitan University 2000 to 2001
Adetutu Aboaba
Alberto Gonzales Benito
Stephen Citrone
Liz Crisp
Veronica Hale
Sharif Herad
Raymond Leung
Angela Silva Jones
Woo Young Song
James Ross
Rozia Adenan
Georg Kaiser

London Metroplitan University 2001 to 2002
Naomi Day
Jean Dumas
Shirin Homann-Saadat
Chris Hale
Robert Johnson
Lole Mate
Paris Sargologous
Thomas Paston Bedingfield
Keith Smith
Magdelena Raczkowska
Jason Wells
Timothy Wong
Ben Brown
Andrew Fortune
Angels Lopez
Namee Im
Tea Puric
Simon Toussaint

Pristina University 2000 and 2001
Shqipe Spahiu
Ardita Byci
Arbenita Ymeri
Eliza Hoxha
Zijadin Hoxha
Besnik Preniqi
Bardha Meka
Merita Quranolli
Armond Gashi
Albana Ramadani
Mimosa Sulejmani
Visar Haxhaj
Pellumb Hulaj
Artan Hapqiu
Venera Haxhaj
Valmira Ajeti
Blendi Spahija
Agron Islam
Gresa Bakalli
Nol Binakaj
Ertan Sulejmani
Burbuqe Latifi
Kreshnik Rraci
Zana Nixha
Armend Fazliu

Oxford Brookes University 2001
Diego Carrillo Messa
Widya Sujana
Erik Lang
Dimitris Ioannidis
Halyna Tataryn

Oxford Brookes University 2002
Lana Al Shami
Nancy Peskett
Colm Dunphy
Jo Fowler
Ayaka Takaki
Keith Daly
Pratima Nimsamer
Viviana Vivanco
Argus Gathorne Hardy
Nate Lielasus
Phaidon Perrakis
Yenny Gunawan

Acknowledgements

Shirin Homann-Saadat:
Burbuque Latifi
Caroline Oakley
Charles Calvert
Dukagijn Hasimja
Farhad Saadat
Floatworks London
Gerda Leopold
Ilir Gjinolli
Jane Wernick Associates
Künstlerhaus Hamburg-Bergedorf
Laurie Anderson
Maurice Mitchell
Mustafa, Vushtrri
Rebecca Horn & class
Robert Frank
School of Architecture & Engineering in Prishtina
Sewing workshop, Vushtrri
Tahereh Esfahani
The boy at Vushtrri's market
Ulla & Heinz Lohmann
Uta Homann
Woman on the bus from Vushtrri to Prishtina
and everybody mentioned in the film 'Dance your shame away'

Foreword

Wherever possible we try to help our students understand and engage with the society they are a part of and will serve. This commitment takes many forms: from working on live projects with real clients locally in London, to international projects in areas subject to intense social deprivation and political change. Such projects allow students to engage with the social, political and economic infrastructures that predetermine built form, empowering them to take a proactive role, generating possibilities and directing their own professional futures.

The work of Maurice Mitchell and his students in Kosovo completed between 2000 and 2003 exemplifies this commitment. The compassion, dedication and generosity of the work points to a new generation of architects who question the elitist values that dominate design culture…architects who are struggling to rediscover architecture as a relevant and accountable social art in the most exacting of contexts.

This is cause for great optimism.

Robert Mull
Head of Architecture and Spatial Design
London Metropolitan University

Chapter 1
Introduction

2

Over the two academic years September 2000 to September 2002 architecture students from North London University (now renamed London Metropolitan University), with help from Development Workshop, have worked alongside students from Pristina University to produce design ideas related to the situation in Kosovo immediately after the end of the war with Serbia.

Our aim was to understand and learn from the situations and experiences of the people we met and to develop an architecture that responds to and evolves with their everyday lives and ambitions for the future.

During the first year we worked in Vushtrri, one of the areas worst hit by the war. An ancient town of about 50,000 people, it lies half way between the capital Pristina and the northern town of Mitrovica, which straddles the border with Serbia. Not only had mosques, schools and clinics been destroyed but so had local records of the location of basic infrastructure such as sewers, water supply and electricity networks. Indeed, very little of the public face of Vushtrri remained. Streets and pavements were churned up and uncared for. There were no squares suitable for the public to meet in and the central mosque had been destroyed and removed stone by stone.

Vushtrri was an important site for us because of the crossover between its strong vernacular traditions and the optimistic energy of its younger generation, who provided the most immediate and vital contact between students and community. In the second year, Pristina was added to our field of study. This larger, more complex, rugged urban space had been relatively undamaged in the war but had since become the centre for a massive aid effort by the international community.

In both communities our language difficulties and more general lack of familiarity with each other's culture were soon dispelled by the hospitality and generosity of everyone we met. Talk was of their emerging independence and new links with Europe and the global economy. In their conflict with the Serbs, Kosovans have relied on their history, language and religion to construct a coherent culture. Perhaps surprisingly, given such a reliance on traditional identity, there was a willingness to embrace civil society within a rights culture.

In Vushtrri we were hosted by Development Workshop, a non-governmental organisation (NGO), who were repairing and constructing new houses to replace those damaged during the war. They provided accommodation, access and guidance. In Pristina, the help of the local university students was invaluable to our understanding of their country's culture and history.

This publication records the theoretical concerns and teaching methods used. It reflects the positive nature of the students' experience and the educational benefit they have derived from it. The unfamiliarity of both the social context and locally observed technologies encouraged experimental (and therefore educative) proposals from the architectural students. The hands-on, site-based nature of building work together with the restricted palette of available materials in Kosovo made the process of building more transparent than most capital intensive construction in the UK. This transparency of process gave the students confidence in the buildability of their proposals. For their part Pristina students were exposed to a wider teaching experience, whilst the families living in the areas studied became involved in an optimistic discourse about their future physical environment and the potential for the development of public spaces in their neighbourhood.

The studio

1. Aims

Design teaching for the postgraduate diploma in architecture at London Metropolitan University is via a studio system. Each studio places a different emphasis on its teaching programme. At the beginning of each academic year students choose the studio they would like to join for that year. The aim of Diploma Studio 6 is to look, listen and uncover new ways of understanding and designing for today's society. More specifically, it seeks to explore and potentially diversify the role that architecture can play in a context of rapid technological advance, where social and cultural changes require an architecture which is immediate, responsive and relevant. While the Studio has tended to focus on locations in the developing world, each year preliminary projects are based on sites and situations located in London.

For many students intent on a meaningful architectural career, questioning the role of the architect is critical – particularly in a world wracked with violence and conflict, where billions of people live in poverty with no access to clean water and electricity or to goods and other services taken for granted in the West. Much of the world lives in urban and rural environments that are unseen, let alone designed, by architects. We face a dilemma... In a world where social justice is not evident, where rights and responsibilities continue to be usurped by national and company loyalties and by unthinking respect for outmoded practices, do we continue to teach architecture as before? Do we continue to reinforce the current trajectory towards environmental degradation? Can we increase the potential for local makers and users of buildings to rediscover the emancipatory experience of building, or are these areas where architects should fear to tread? We do not wish to abandon our involvement with form- and space-making, but to enrich and deepen it with a wider environmental, social, cultural, technological and developmental understanding.

Whilst the studio attempts to bridge the gap between development discourse and architectural practice, it also questions the characteristic commodification and consumption of architecture production today. Studio teaching encourages a healthy distrust of the heroic and the fashionable and seeks inspiration in the everyday lived experiences of ordinary people. Numerous social, spatial and aesthetic meanings can be found in the activities and conditions that constitute our daily, weekly, seasonal and yearly routines. The ordinary stuff of everyday life reveals a fabric of space and time defined by a complex realm of social practices: a conjuncture of accident, desire and habit (Crawford, 1999). This notion of the everyday is particularly important in a context of rapid change, where the ability to carry on with one's usual daily life can signal the shift from a state of emergency to that with a semblance of normality.

The Studio's investigation of the 'everyday' involves the reuse and adaptation of the urban landscape. More particularly, architectural insights are extended beyond those of space to include specifically those of time. Within Studio 6 design teaching emphasises the refinement and continuous reshaping of landscape and buildings. Students are encouraged to examine how best to reuse and recycle hitherto redundant structures and materials and to design so that they can be adapted and reused in the future. Studio programmes usually include an 'on-site exercise' where students work with users to design and construct something of immediate impact and relevance for the community. In these exercises full size experimentation with locally available materials and construction techniques are employed as an investigative tool to drive the design forward.

2. Themes

The Studio's areas of interest have grown out of a number of cultural and technical themes emerging over the last 30 or 40 years, most of which have not, in the past, been central to architectural thinking but which are considered by the Studio teachers to be relevant to its aims. These are listed below: a bibliography of relevant texts is given in the last section of the book.

Cultures of making

Students are encouraged to get involved with the process of making buildings on-site, and to value the opportunity this affords for the creative and imaginative input of local makers to the building design. Making is seen as a performance that connects with what was there before and with other time specific building performances in the future. We are interested in the differences in the location and quality of creative input dependent on whether buildings and their components are constructed on- or off-site. In this regard study of social mechanisms such as self-help co-operatives and cash minimisation schemes, which release hidden resources of imagination and sweat equity from cash poor cultures, are encouraged.

Rights cultures
Within the projects, assumptions are made that change towards a situation where positive rights to shelter, health, education, clean water and safety and an awareness of women's rights is required for the effective participation of responsible individuals in a functioning civil society.

Cultures of situation
Situationist methods (explained more fully in Chapter 2, Teaching Tools and Methods) are used to record, study and make a series of small but pertinent design interventions which affect and improve the user's experience of everyday life.

Vernacular architecture
Vernacular architecture is often used by the students to exemplify the direct expression of lived experience. At the same time they are careful to avoid the cultural baggage associated with nostalgia in the conservation of traditional structures: the reactionary fossilisation of habitual ways that inhibit change and resist searching questions, favouring legitimacy and orthodoxy over experimentation.

Appropriate technologies
A range of alternative or intermediate technologies, which has emerged over the last 40 years from the discourse on being appropriate for a specific context, are used to help integrate technological experimentation by the student with site based studies in Kosovo. These include an interest in the effect of climate and energy on building design; sustainable services; loose-fit technologies; autonomous houses; temporary and mobile buildings; buildings in use; renovation and repair; phasing; disaster resistant construction; self-build construction; green construction.

Out of a fascination with these cultures and technologies, the Studio staff have put together a series of teaching methods outlined in Chapter 2 intended to help the student understand the context and design accordingly.

3. Relevance

What role can the study of architecture play in a postwar situation?
Quoted in an evaluation report prepared for the British Council on their activities in support of the University of Pristina, an architectural student said that: 'The best thing that happened was the coming of the North London students…the way they learn things is very different.'

During the second year's field study students from London and Pristina worked and played together intensively for a week and had endless discussions and friendly arguments about alternative ways of producing buildings. The Pristina students' architectural education focuses on design using mathematics and engineering whilst UK students emphasised drawing and the recording of social interaction. Frank exchanges of views were solicited and encouraged. Differences of opinion were valued.

War, and natural disaster, usually mean that new strategies need to be put into place to cope with emergencies that the old order has failed to avoid or foresee and has manifestly failed to provide for. Thus designing and building in a context of rapid change often challenges the legitimacy of old methods and opens up metaphorical and physical spaces for new meanings and functions to take hold. Objects and spaces may have lost their original function and, with that loss, meaning is often also displaced. Do we respond with reaction or resistance, or do we embrace change, unclear of its direction? There is never only one way, there is always a choice. Ways forward guided by general concepts like 'appropriateness' or 'sustainability' are never adequate on their own. In the field of architecture, we need to discover, construct and make explicit new social and material relationships from the leftovers of the past and people's ambitions for the future.

New validity must be sought by probing the imagination – with the process of making, both socially and materially – and in an understanding of the lived experience of the people. For those involved this can be both an emancipation and a challenge. Space is always saturated with qualities that architects can explicitly represent in order to flag up the dreams of society. In a situation such as Kosovo, living under what Adorno (Leach, 1997) has called 'the shadow of instability', these images need to be upbeat and generous.

Such optimistic images are usually absent from the reconstruction and development effort that concentrates, quite rightly, on the provision of basic 'no frills' infrastructure. But if the opportunities inherent in new construction are to be realised then newly emergent fragments of social and material endeavour need to be recognised and supported.

Often, because of the hands on nature of their work, smaller organisations working on the ground, like Development Workshop, are more likely to be able to engage and make explicit energising images which can boost local initiative. However, the aid industry has modified its 'development' paradigm within the built environment. Finding it increasingly difficult to cope with the commitment implicit in development projects, northern NGOs and national aid agencies have, of late, concentrated on 'disaster management'. This prioritises a 'quick in, quick out' approach to the neglect of long term involvement with on-going improvement. Because of the need for speed: for highly visible, rapid results, interventions within the physical environment have tended to be standardised products where the creativity and invention embedded within the product is highly remote from its application. Doctors and nurses have supervised the construction of standardised clinics, whilst teachers build textbook schools.

Such standardisation sidesteps the importance of embedding a dialogue with the users within the process of change. Dialogues help to promote familiarity and eventually the adoption and 'ownership' of their built environment by the users. Elsewhere (Mitchell, 1995) I have argued that there is a gap in architectural education in the UK for those wishing to specialise in the skills and techniques necessary to support a sustainable process of building production: that is within the constraints common to developing countries, where design is generated by specifically local conditions and at a level where both the makers and the users have a measure of control. Studio 6 aims to work towards plugging this gap.

A Kosovan house wrecked through looting

Kosovo

1. War

Kosovo is a society in transition. For three months in 1999 NATO carried out its largest ever combat operation there, with the aim of stopping Serbian oppression of the majority ethnic Albanians. On 10th June, after a sustained 78 day air offensive, Yugoslav President Slobodan Milosevic agreed to withdraw 40,000 Yugolsav and Serbian troops from the province leaving the way clear for the international peacekeepers to move in.

NATO bombing, Serbian military operations, looting and ethnic 'cleansing' had wrecked more than 120,000 buildings of which half were homes. Half a million people were left without shelter. At the time of our first visit 16,000 houses and 80 per cent of all schools had been cleared of mines and 15,000 mines and cluster bombs had been cleared from public areas.

Following NATO's intervention in Kosovo and the formation of the United Nations Interim Administration for Kosovo (UNMIK) the European Union (EU) funded various NGOs to work with local people to rehabilitate houses for 20,000 refugees. This included Development Workshop (DW) which was asked to help with the rebuilding of 300 houses in Vushtrri.

A KFOR *military convoy, 2001*

Rioting in a Kosovan housing estate

2. The public realm

In the first year of our investigations the United Nations had limited the scope of their funded projects in Kosovo to housing repair and they were pushing for a relatively rapid hand-over to elected representatives. However, to function effectively, modern civil society needs not only good governance but also sustainable public spaces, schools, healthcare and recreational facilities, shops and markets, transport, sewers and garbage disposal, clean water and power supplies (Rowe and Sarkis, 1998). Thus the teaching programme was concerned with the role that architectural ideas could play in the construction of a new public realm in urban Kosovo.

In the first year after the cessation of hostilities there were street demonstrations demanding the release of prisoners from Serbia. By the second year these had given way to a heady mix of optimism in the streets, cafés and bars. The ancient, sometimes imaginary, past was a touchstone of stability for the older surviving population, yet by 2001 the unprecedented cosmopolitan mix of aid workers, returned intellectuals, internally displaced people and vibrantly energetic young adults were all working for a modern Kosovo joined to Europe.

In 2001 the urban landscape of Kosovo was still damaged, its water and power supplies only slowly improving and its major historic and public buildings still missing. The mood of the people, however, was less disillusioned and the scene was set for the re-inhabitation of the postwar urban landscape with the built structure of a more optimistic civil society. The ambition of the Studio was to explore the potential for sustainable loose-fit technologies to enable Kosovans to edit, reconstruct and recycle what was already there and add less familiar elements, where appropriate, in a continuing process of experimentation, invention and self discovery.

3. Field work and investigation

The students visited Kosovo in November 2000 and 2001. Each field trip lasted about two weeks. Students were divided into groups of about four. Each group measured-up and generally surveyed a particular area and carried out 'On-Site Exercises' to become more familiar with its inhabitants. In the first year, in Vushtrri, five areas were investigated; in the second year three 'Areas of Investigation' were selected in both Vushtrri and Pristina.

Back in London each student constructed a design programme from the experiences recorded during the field trip. They used techniques of narrative and metaphor to attach identity and meaning to optimistic scenarios of change which culminated first in a design idea and then in a resolved design project. Student schemes ranged from programmes for self-sufficient communities to therapeutic spaces and from public outdoor spaces and post offices to educational buildings.

The technologies chosen by the students for their projects was integral to the social and cultural meaning developed from their scenarios and often played a large part in the construction of their design ideas. The process of integration was clarified by experimental work carried out at the Centre for Alternative Technology on the students' return from Kosovo.

An illustration

Postwar architecture in Beirut: an engagement with the landscape

By the time the two opposing sides stopped shooting at each other in the streets of Beirut in 1975, a swathe of land had been carved out by the gunfire. It ran through the heart of the city and was named, perhaps ironically, 'the green line'. The terraced buildings edging this space, peppered with shell and bullet holes, were unusable. The space between, initially clogged with rubble, went unclaimed by either side and as such was one of the few spaces in which citizens from either side could walk their dogs in peace. Once the rubble was cleared and grass had taken root it became the only truly municipal, linear 'green' park in the city. This park was constructed from both the mental and physical spaces opened up by the conflict. Its validity was recognised, represented and sold to the opposing sides by the architects and planners who tidied it up.

Students surveying with the help of school children

Vushtrri's traditional Turkish Hammam in the market square

Chapter 2
Teaching tools and methods

Kosovan children playing before the war

Strategic physical and policy planning has a tendency, because of its large scale and protracted implementation periods, to ignore, blight or at least undervalue the rich, local, small-scale, rapid interplay from which a diverse, creative and convivial society is constructed. In so doing it often fails to take account of the spatial, temporal and environmental conditions within which these exchanges take place. For this reason the studio has encouraged students to work from observed fragments of lived experience (called 'moments') outwards. In order to expose rich local interactions, issues and concerns, which are usually silent within planning discourse, the Studio needed to find techniques to explore the urban fabric in a creative manner.

Each element of the everyday is related to the next in space and time. It is an animated experience for the participant. It reflects something specific about the place and time that it is situated within and those who initiated it.

1. The Situationist Derive

To facilitate this way of working the studio adapted the technique of the Situationist Derive. The Situationist movement was a response to the belief that modern cities had reduced their occupants to being mere spectators of life without taking part in or involving themselves in the events that were taking place around them. The Derive is a tool for mapping and plotting areas of activity, movement and encounter. The mapping does not reflect the normal two-dimensional street atlas, but instead shows in a graphic way our emotional responses to an area, its characters, a perception of its people and the recording and identifying of its zones. It is a tool designed to navigate you playfully through a space detached from any conscious decision about choice of direction of movement, making you focus instead on how you feel, whom you meet and what you experience. The students practised this technique in a small preliminary project at the start of the year and then adapted it for use in Kosovo.

2. Site survey

All students, working in groups of three or four, were required to carry out a large-scale measured survey of their chosen Area of Investigation within which they would eventually each choose an individual site for their design. These Areas of Investigation varied in size according to size of the student group and the physical boundaries observed. For example, 'The River Group' looked at the locality related to a small river running diagonally through Vushtrri, whilst 'The Ashkali Area' was defined by the buildings surrounding an open piece of common land, which the students named 'The Gap', and which had previously been occupied by an ethnic group called the Ashkalis. They would accurately record spaces and materials and beyond that the occupation of the site and something of the changes from day to night, say, or the changes required on a market day.

3. Moments

Within the student's Derive, particular small events such as a snatch of conversation or the process of alighting from a bus, for example,were recorded in their spatial and temporal context, using drawings, photographs, video and tape recording techniques. These events were called 'moments'. The particularity of each moment was considered, not just its location in space and time but through a sensory three-dimensional mapping of the space and those who passed through it. For example, a moment might record what you could hear, see and smell, who you met, what they said, what they were doing and their use of space. These moments were then used by the students as a site, a time and an event (a situation) from which to develop briefs for their proposals.

This method of working allowed the students to research a site and identify a user/client uninhibited by conscious restraints. The level of interaction required during this process allowed a glimpse of the patterns and complexities of everyday life at the site and enabled the students to develop an interpretative vocabulary with which to communicate these explorations. This in turn led to the development of architectural proposals that responded imaginatively and optimistically to the needs and desires of individuals and families mostly within the public realm.

4. On-site projects

In addition to borrowing and adapting Situationist techniques in order to record their own personal impressions, the students found ways to communicate and interact with people inhabiting and using their Areas of Investigation. The most curious and approachable members of the local community tended to be children. They became fascinated as they looked over the shoulders of students sketching, surveying and photographing their familiar environment. They often challenged the students' representations substituting their own impressive sketches indicating different preconceptions and priorities. One of the most strikingly successful examples of an on-site project was Andrew Fortune's child-constructed, 24-hour, plastic bottle bridge jamboree. Through the collective efforts of himself and the children he was able to highlight the potential of a rubbish strewn watercourse.

Vushtrri inhabitants demanding the return of their men from detention

5. Water

Themes are offered to the students as a way of focusing their minds on issues relevant to the year's programme. They provide an opportunity for students to acquire 'another way in' to designing. They are best left general enough to combine technical and aesthetic attributes: the physical and the poetic. The extent to which a theme is picked up depends on the student's own path through the design process. For both years of study, the role of water in the making and experience of public space was explored.

In Vushtrri a sustainable clean water supply was still lacking, whilst demand was steadily increasing. Between the Studio's first and second visits Kosovans had taken to car and street washing, using high pressure pumps and sprays attached to the mains water supply or to rainwater storage tanks. This was a very noticeable street activity. Ideas about rainwater collection, filtration, storage and distribution, its use for washing, bathing and cooking, together with sustainable treatment of waste and sewerage using reed beds were included in many students' programmes. The potential contribution of water to an emergent public realm of leisure and sport was also explored.

Swimming pools were high on the Local Authority (UNMIK) wish list for new public spaces. Chris Hale found that safe, water-based play was the major theme emerging from his site enquiries along the river in Vushtrri. As a result, his proposals integrated a public pool within the river bank, with a corresponding scheme for allotments and reed beds.

Another student, Shirin Homann-Saadat, emphasised the poetic qualities of water. In her on-site exercise in the market square she used a puddle to represent a fountain in front of an old redundant hammam, suggesting calm and relaxation. The regular drip of water gave a sense of place to the area within which the sound of the fountain could be heard. This metaphor of water-borne calm was carried forward into her finished proposal where a flotation tank provided a sense of stillness and healing for traumatised women.

Kosovan children urging a game of football on the students

6. Narratives

A story provides a framework for understanding experience. A narrative, in itself, is neither true nor untrue but, like a novel or a piece of music can contain meaning and, suitably represented, can communicate the kernel of a design idea. Validity and meaning in an architectural narrative is embedded in the information chosen to represent a series of moments and can be constructed from a continually changing and unfinished engagement with the environment. Students are encouraged to appropriate narratives from the fragmentary situations they encounter as an aid to developing a design idea.

Narratives usually contain repeated elements, each repetition being slightly adjusted to adapt to an evolving story line. James Ross used the idea of playing a game of chess (a national game encouraged by the Local Authority) to provide a focus for a number of scenes within his intervention. This led to a tweaking of his more literal programme (a sustainable market gardening community) to provide space and opportunity for increasingly formal games as the community developed.

In design teaching it has been normal practice to talk about resolving design problems. Narratives, however, are more concerned with viewing changes over a bracketed time period. Storytellers choose the position of the brackets and edit the content to display a sequence of scenes evoking a pattern of tension and release. Naomi Day's story of the construction of a swimming pool, with minimal local resources, started by representing the tense position of returning refugees unsure about what they would find at their entry point into postwar Vushtrri (an anxiety mirrored in the tension felt by a diver before take off). Would their house have been destroyed? Would they now be accepted by their new neighbours? The narrative transformed this downbeat scenario into a wave of public optimism about sending a very visibly well-trained diving team to the Olympics, signifying Kosovo's emergent status as a modern society within the global economy.

7. Metaphor and naming

A metaphor comprising a system of names or labels can transfer the essential meaning of one set of circumstances (the observed moment) to another (the design idea) without having to reduce its complexity. For example, in the Gilbert Islands the conflict of male and female within a household is expressed by naming opposing rafters on either side of the roof ridge after family members. Just as the roof could only be supported by both sets of rafters, so the society could only flourish with the co-operation of both men and women.

In developing a design idea the rich assembly of named habits, which comprise a metaphor in the original context, provide an organising tool (after suitable adaptation) when transposed to another newly framed situation. Carried on the back of the organising function of the metaphor is the cultural baggage of its previous incarnation, which will itself need editing and tweaking to make it fit in its new place. The metaphor is particularly useful as a demonstration of neatly encapsulated meaning and as a means of communication with the user.

After a week's study Liz Crisp was able to plot the movements of a family squatting in a damaged house among the ruins of the Ashkali Area of Vushtrri. She named the spaces they used in accordance with her observations. She used this knowledge as an organising metaphor for her scheme. For example, notions of the veranda as a threshold, offering a secure place for the family to hang out and communicate with passers by, and the recycling of bricks from the rubble, were combined in her design for a gabion plinth for the veranda outside her internet café.

8. Re-engagement

At the end of each field trip a meeting was held of all the people who had taken part in the exercise (Pristina and London students, Development Workshop, Local Authority and local inhabitants). The purpose of this meeting was to show the photographs taken, and to explain how the students would use their accumulated information and experience once they got back to London. Small gifts were given to the community as a way of thanking them for their co-operation. The first year's meeting became the subject of a radio programme hosted by Ade Adegboye. A football match between the older children and some of the students closed the proceedings.

During both years students from Pristina School of Architecture worked with the North London students, helping to identify the Areas of Investigation, undertaking introductions to occupants and in the second year undertaking parallel design projects.

An exhibition of the first year's design projects was held at Pristina University at the start of the second year's field trip. This was attended by the press and broadcast on television. A presentation of the work was also held at the UNMIK municipal offices in Vushtrri. Halfway through the second year, students and staff from Pristina visited London and attended critical presentations of the London students' work in progress, at the university, allowing a full debate over issues arising from the work.

9. Large-scale modelling

Architectural design usually operates at some distance from the making process. In order to re-establish a closer relationship between design and making, the field trips to Kosovo were each followed by a four-day modelling workshop at the Centre for Alternative Technology in Wales. These workshops gave the students the opportunity to experiment with unfamiliar materials and to investigate the spatial impact of unfamiliar technologies on the landscape. For example, one group of students constructed a large scale site model of the difficult terrain at the edge of Pristina. This provided spatial insights key to the development of their design ideas.

The precise amount and type of technical knowledge in each project is also difficult to determine in advance. For the modelling process to contribute to the architectural intention it should be exploratory. Experimentation with unfamiliar materials is preferred to recreating known technologies because of the restricting nature of their accompanying baggage. Conventional technologies embody such a strong nostalgia that they inhibit searching questions, favour legitimacy and orthodoxy and deny the educational value of experimentation in all but the most rigorously controlled laboratory conditions. It was something of an advantage to the students that they were operating in a situation where available building technologies could not be taken for granted.

It was in this workshop that Naomi Day constructed her views over water (key to her design idea) and Andrew Fortune experimented with earth bricks and bottle walls. Namee Im constructed her half-scale horse shelter from polythene and scrap timber and Jean Dumas explored different retrofit cladding ideas.

10. The design idea

By the end of the first term students were asked to come up with a design idea which might form the basis of the resolution of the intervention to be represented in portfolio at the end of the year. The techniques described above promote the ongoing process of social interaction with the environment as the generator of meaning. The work produced tends to reflect the spontaneity of this way of working, which by its very nature is difficult to resolve. However, the student experience is represented as a single portfolio performance wherein the changing needs of the users are captured, framed and represented as architecture.

The design idea is an organiser of space, narrative and meaning. It needs to have the immediacy of a visual image and the rich complexity of a metaphor. A good design idea will be sufficiently robust as to be able to absorb changes in the programme as the design develops, without losing its resonance as an organising idea. Often design ideas can be represented by a key image and a name. Naomi Day's 'View Over Water With Diver' and Magda Raczkowska's 'Community Canopy' are examples of this approach. In other cases the names of iconic structures available for occupation on the site are combined with programmes developed from existing activities: a post office in the 'Domino House' (Vera Hale) for example or a theatre in the 'Warehouse' (Jean Dumas). Lole Mate and Tim Wong developed ideas from a realisation of the potential of their site for solar power generation and the spatial impact of that transformation on its cliff-like landscape. Raymond Leung's idea for Castle Square was derived from a view over the town from the space between an ironic juxtaposition of squatted tower block and crumbling castle.

Chapter 3

Narrative architecture: The Vushtrri Red Black Box

by Shirin Homann-Saadat

1. Towards a narrative architechture

The architect was once defined as someone who knows the reasons for his or her actions; it was claimed that the craftsman knows the 'how', but that the architect also knows the 'why' of a project.

If architects know 'why' they are doing what they are doing, they also know the reasons and motivations behind their work. Sometimes, however, these motivations seem well hidden. Explanations of why projects are done in certain ways are closely connected to economic issues. Therefore architects more frequently ask how they work within a particular set of rules, rather than asking why they work the way they do.

The rule-led aspect of the architectural profession goes well beyond economic constraints: we agree to a fixed set of means of representation. We respect building laws, even if they are outdated or harmful to the environment. We study an increasingly person- rather than work-driven architectural press. And it is inherent in our profession to accept and nourish working conditions that shrink our lives to such a degree that it echoes back into the architecture we create. Just like other professionals we accept rules, which we might criticise privately, just so that we can be part of the profession. Sometimes, though, we face projects that blur the boundaries between our professional and personal rules and values. And I suspect that it is this total challenge that makes us reconsider why we work the way we do, personally as well as professionally. To me, working in Kosovo is such a project.

Intense conversations with women, children and architects from Kosovo made me believe that architecture should address not only the spatial but also the narrative details of a site. Their particular stories proved that a problem-based programme is just as important to a project as the widely accepted celebration of architectural shapes, technical details and materiality. If we don't look towards a narrative architecture we run the danger of getting lost in pure aesthetics, drowning in visual data – an experience akin to living with a beautiful person who has little to say.

The Kosovo stories I was told were a reminder that our responsibilities and possibilities as architects go further than we tend to think. Attempting to address politics and ethics through our work does not mean leaving but rather extending the field of architecture. Any architecture – small- or large-scale – is a three dimensional ethical expression, defining and reflecting values that shape and mirror the way we live, but such expressions need to be achieved consciously not accidentally, if we are to act responsibly in our lives and work.

The concept of a narrative architecture is drawn from the stories inherent in a site: the intention is to add to those stories as well as to the built environment. As an attempt to develop an architectural language it will hopefully make it possible to answer not only how but also why a project is approached in certain ways. It is also a

a) Long after the war Kosovo is still dependent on generators due to regular power cuts

b) Vushtrri's Ashkali area in 'peace time' (2001)

c) Mental map of Vushtrri

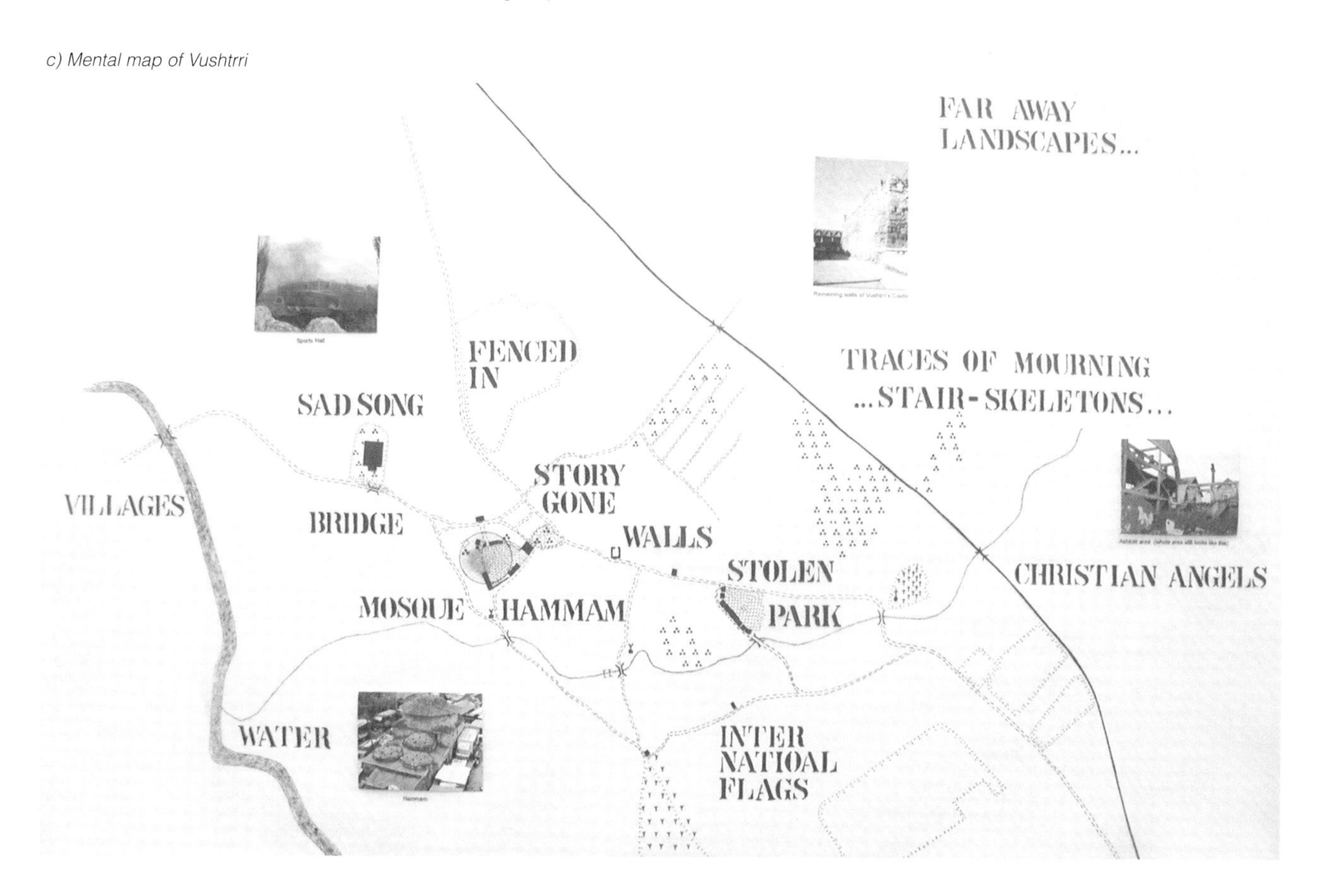

way of saying 'no' to the widely accepted view of architecture, which allows remarkably little space for those aspects of life that cannot be seen or touched. I am thankful to the people in Kosovo for sharing some of their stories and reminding me of the fragile, invisible landscapes out there that are important to us no matter where we come from.

2. Mental mapping of Vushtrri

Vushtrri is a town on the main road between Pristina and Mitrovica. The state of its infrastructure makes one wonder about the quality of the 'peace' that people claimed to have brought to Kosovo. With the advent of so-called 'preventive wars' one can assume that the definition of 'peace' and 'peace bringer' will decay even further.

In Vushtrri, long after the war's end, the water and electricity supply is still poor (*a*,opposite). Many of its buildings are far away from resembling sheltering structures and proper roads are almost non-existent (*b*,opposite). Nevertheless, a huge part of Vushtrri's social life happens outside in the public realm, in the not-yet-rebuilt-streets of the town. This means that many activities take place on either dusty or muddy ground, depending on prevailing weather conditions.

Drawing a mental map (*c*,opposite) of the town's ground conditions, two patches clearly stood out: a small park at the north eastern edge of Vushtrri and the market place in the centre of town. While the park is a soft, green patch on the map, the market place is one of the rare asphalt covered public realms. Both places are examples of the few urban carpets within Vushtrri that provide 'safe ground' in a sea of work-in-progress. The park was once used by Vushtrri's children. These days they claim, they cannot go there anymore because 'the grown ups have taken it over.' The market place, on the other hand, is primarily used once a week, on Friday – market day in Vushtrri. During the rest of the week it is rather deserted, with only a few children cycling around or playing games.

Every time I went to the market I saw the little boy who had initially told me that the grown ups had taken over their park. He described the new games he had invented and how he wanted to play them one day. This little boy and his dreams made me wonder how one could make his 'asphalt playground' more appropriate for his imagined games...

South-west view of Vushtrri's market square (empty patch of land = site for the project for women)

Above: The boy who invented many games in front of the long wall and site, Vushtrri's market
Right: Painting the town workshop/ turning a puddle into a wishing well, Vushtrri's market

3. Using the ground as canvas: painting the town
To improve the boy's asphalt playground something had to be achieved on the spot with very few resources. It was important to make changes quickly, rather than turning this ambition into a major project that might only be realised by the time the boy was a grown-up. Therefore the decision was taken to go back to the market with some paint and brushes: to 'paint' this part of the town, rather than sit in front of a computer or drawing board planning the intended improvement.

We started by painting around a huge puddle on the asphalt and adding some Albanian words to it, to indicate that this puddle could be read as a wishing well. As we continued to paint games on the ground, local children as well as grown-ups began to turn up and ask if they could help. First they brushed clean every patch we had decided to paint on and finally they started to paint and write themselves. What began as an idea to reanimate the underused asphalt ground at Vushtrri's centre had turned into collaboration between people from Vushtrri and London.

To some the results of our work might seem to be little more than graffiti , but there is no doubt that the painted bits and pieces were direct expressions of the residents' dreams, needs and beliefs. In addition Vushtrri's children got temporary, self-made playing fields right in the middle of their town.

Left: Painting the town workshop/ youngsters & grown-ups joining in, Vushtrri's market
Right: Turning the long wall into a wailing wall, Vushtrri's market

No doubt, painting a public square is in some way illegal. At the time, one participant expressed concerns that 'the UN might not be happy with us painting the town.' But since a lot of Vushtrri's inhabitants joined in and since it is their town, one could argue that it is sometimes more appropriate to ignore rules protecting an undesirable status quo.

Six months later a colleague from the School of Architecture and Engineering in Pristina presented her latest project to us. Burbuqe Latifi had designed a redevelopment scheme for Vushtrri's market in which she had 'officially' incorporated playing fields into the market place, exactly where our workshop had taken place.

4. Turning the narrative context into an architectural programme

On a bus, somewhere between Vushtrri and Pristina, a woman told me about her three daughters who had been raped during the war, two of them right in front of her eyes. The woman did not blame any single individual but said: 'It is in our culture to feel guilty about it.' When I asked how they were coping she replied that they could not cope at all. She mentioned that there were international agencies for women in Pristina but pointed out that most of the women she knew would not go to them. She said that there were still too many other problems in Kosovo as a whole to solve for them to be able to concentrate on their 'individual problems'... 'The real problems will start when Kosovo's general problems are solved.'

The woman seemed reticent, though she was the one who had started the conversation. When I left the bus I asked her why she had told me this seemingly very private story and she replied: 'Because it is important that other people know about it – probably that will help.'

A bit later, as I was walking through a part of Vushtrri that is still devastated, I saw three women sewing in one of the less destroyed buildings and they asked me to come inside to join them for tea. The women had turned their space into a small workshop and shared one heater that was constantly pushed from one sewing machine to the next. Their work was delicate and beautiful, in complete opposition to the conditions they were working in, and it made me think of the words of the lady on the bus: 'It is important that other people know about it – probably that will help.'

Left: Vushtrri's sewing workshop
Right: Butterflies dancing in space/ Vushtrri Dance School

On my way back to Vushtrri's centre I heard music coming from one of the buildings. I followed the tunes and ended up in a room full of musicians and dancing girls and boys. The girls were waving around colourful pieces of cloth as they moved through the space. It felt like being in a room full of butterflies right in the middle of a torn town, a town that was also generating stories like those of the woman on the bus and the sewing workshop in a half-destroyed building. This strange mix of intense impressions made me decide to work on a project for women. I wanted to do more than 'just' plan a building. I wanted to tell the story of what I had seen and heard to as many people as possible and I determined to come up with a design that could be built by Vushtrri's inhabitants.

The idea of making a women's space was obviously initiated by the conversation about the rapes. Since I regard rape as the most private space invasion possible, i.e. that of one's own body, the provision of a sheltered space exclusively for women could be one step in a process to deal with a kind of violence that reaches far beyond warfare. (In summer 2002 the British Government published a study stating that every twentieth woman in Britain has been raped). But since the lady on the bus had mentioned that hardly any women sought help from organisations where they were expected to talk about rape, the 'construction' of another rape crisis centre would have been a pointless idea.

Respecting the women's stated desire to remain more or less silent and 'just look ahead' led to the design of a sequence of room-sized fragments with different uses, rather than a monolithic, mono-functional building. The sequence of protected rooms is intended to deal with the body, education and work, to provide spatially for individually paced healing rather than organised therapy. The project's fragments consist of a 'flotation' room, a 'room of motion' and a small 'tower-to-look-to-the-future'.

The flotation space is a room-sized quote from Vushtrri's existing hammam. Sunk into the ground, the space keeps noise intrusion to a minimum and remains at a constant temperature, similar to the protective qualities of caves. Inside the room is a salt-water bath. Originally developed to cure back pains and nervous disorders, flotation tanks are used to treat many ailments, but especially stress related ones. The salt-water mix makes you float, so that you do not even feel your own body weight anymore. With its cave-like qualities the sunken flotation room is simply a place where Vushtrri's women can keep their sense perceptions to an absolute minimum, providing them with a space that offers optimum physical and mental relaxation and calm.

The second fragment, the room of motion, 'leans' against an existing building, saving on building materials and allowing the new project to work within the architectural context of the site. It

Top left & middle: Tower, flotation room and room of motion, stills from the film 'Dance your shame away'
Bottom left: Construction manual sketch, Step 5: earth bag construction of dome

provides space for activities with the opposite qualities to those of floating. Inspired by the musicians and dancing girls who occupied every corner of their space in a creative way, the room of motion is a place for every kind of movement – including sound. Since rape not only instils silence and shame, but also provokes various forms of aggression, the room houses a small punch-bag forest. The room of motion is the only space in the building sequence that leads directly back onto Vushtrri's market. When its doors are open, the room of motion will become a small performance space, like a 'stage-in-a-box'. The audience, though, will always remain outside the room of motion, on public ground, in Vushtrri's market.

The third element, the tower-to-look-to-the-future, is the entrance fragment of the building sequence. Its ground floor elevations are kept open to the inner courtyard of the project. The first floor would house Vushtrri's sewing workshop with its windows being at sitting and working height, overlooking the courtyard. The use of the second floor is to be decided on by Vushtrri's women. Since many libraries were destroyed during the war, the third floor is intended to house a book collection (women assembling their own book collections is generally important). The roof-garden on top of the tower consists of a poppy field where the women can sit – physically above their day to day life – and read.

The whole building sequence is surrounded and protected by a rammed earth wall that is constructed from the soil gained from the digging out of the flotation room. This wall turns the space between the three fragments into a sheltered courtyard – a smaller second courtyard, in which various herbs can be planted and harvested, sits in front of the sunken flotation room. The three building elements of the project can be constructed step-by-step or in one go. Written like a recipe book, there is a construction manual explaining the various building phases to non-professionals. Together with other items that are related to the project, the manual forms part of the 'Vushtrri Red Black Box', which is going to be brought back to the journey's starting point: Vushtrri, Kosovo.

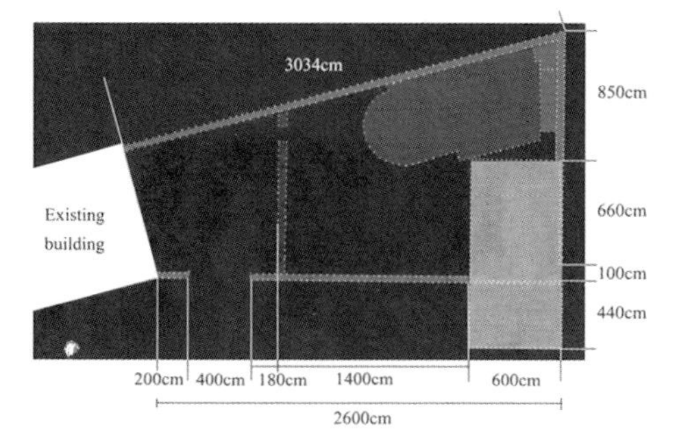

Construction Manual, Step 1: Making a plan in 1:1 (drawing the plan on the ground)

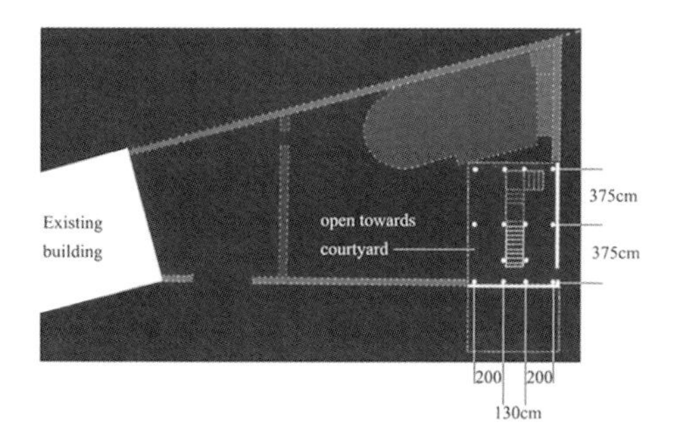

Construction Manual, Step 2: Adding goods to the plan (constructing the tower)

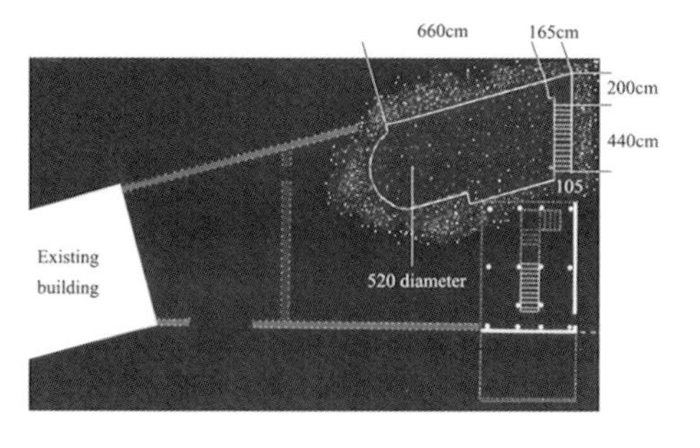

Construction Manual, Step 3: Engaging more ground (flotation room digging)

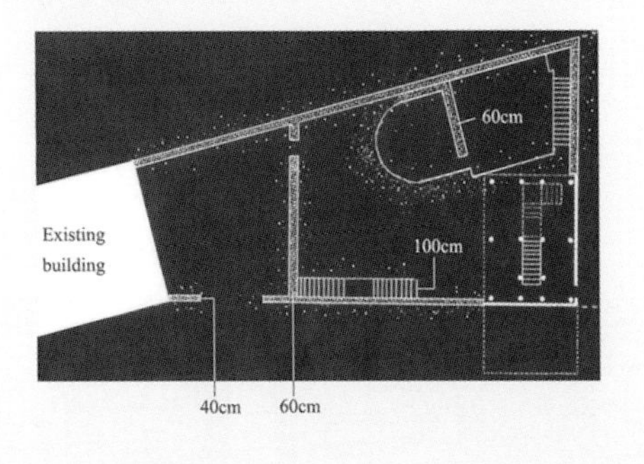

Construction Manual, Step 4: Tidying up as a way of construction (recycling left-over soil for rammed earth walls)

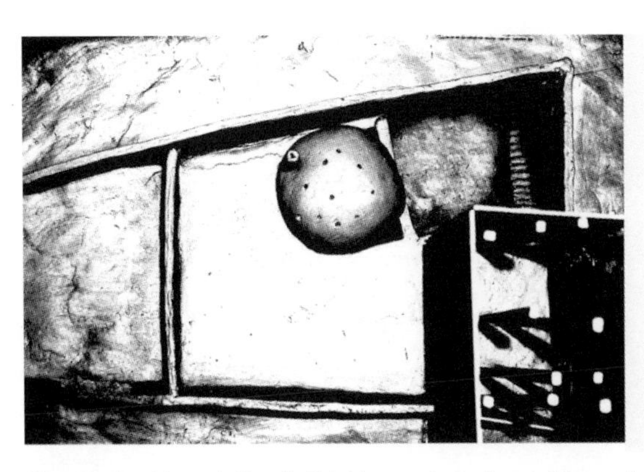

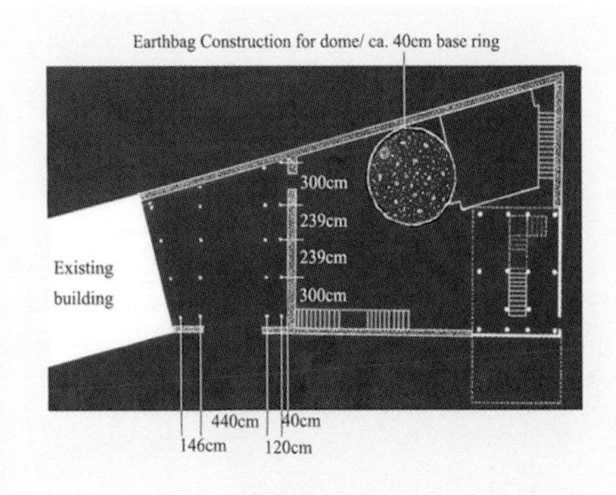

Construction Manual, Step 5: Patching up (recycling soil for earth bag construction of dome/ flotation room)

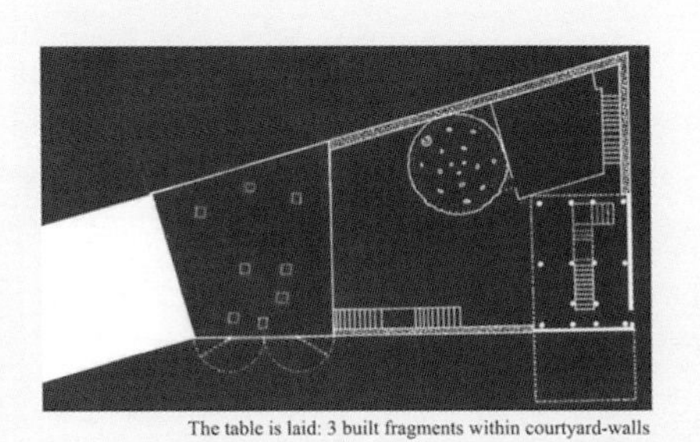

The table is laid: 3 built fragments within courtyard-walls

Construction Manual, Step 6: The table is laid (a building for women in three fragments)

Above: The Vushtrri Red Black Box, closed
Above right: Box open, layer 1
Middle right: Box open, layer 2
Below right: Box open layer 3

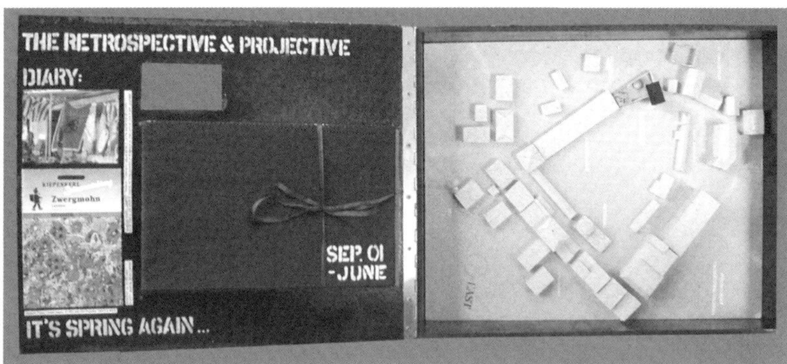

5. The Vushtrri Red Black Box
(stories and architecture in a travelling box)

The Vushtrri Red Black Box is a present to the women in Vushtrri. As a 'black box' it contains data collected during my journey from London, via Kosovo, to the International Criminal Tribunal for the Former Yugoslavia in The Hague.

An attempt at an inverted Pandora's Box it contains:

- the building proposal for the women of Vushtrri (in the form of a film)
- a documentary of the collaborative workshop Painting the Town (film)
- a recipe book/construction manual for the building sequence
- a 1:500 model of Vushtrri's market square and the proposed building
- books for the new book collection
- drawing and sewing material for the tower's first floor programme
- poppy seeds for the roof-garden of the tower
- a guest book to document the box's journeys for the people in Vushtrri and to gather input for the continuation of the project.

Building a box to include the architectural proposal, objects related to its programme and a documentation of the project's journey had various motivations. I wanted to have something that I could bring back to the people in Vushtrri, rather than ending with an academic exercise. It also felt inappropriate to go back to Kosovo with a set of drawings that would be of interest only to architects or engineers. By presenting the design in the form of film I hoped to address a wider audience and to give the proposal a feeling of being less fixed, of being open for discussion.

The documentation of the workshop Painting the Town is included in the box as a reminder that a small-scale collaboration has already taken place and that a more ambitious collaboration – construction of the proposal – might actually be possible. The various objects – books, sewing material, poppy seeds etc. – are part of the box to strengthen the idea that a realisation of the building sequence is conceivable.

A further reason for building a box to 'house' the project was that its size (46cm/37cm/13cm) forced me – spatially – to condense and edit what I had seen, heard and designed. Making the trip to Kosovo, the return to the so-called civilised world (London) and especially the visit to the International Criminal Tribunal for the former Yugoslavia in The Hague, I had overestimated my capacity. At one point, it felt as if I was sitting in a Camera Obscura gathering together all the atrocities ever committed on our planet. Building the Vushtrri Red Black Box, and packing it over and over again for its journey back to Vushtrri, was like building a secret, small door out of that Camera Obscura.

Originally, the box's journey was supposed to take it from London straight to Kosovo. Although it left England in July 2002, it has not yet arrived back in Vushtrri. But the box's excursions and openings have introduced Vushtrri to Berlin, Vienna and recently to the 'Architectural Summer' in Hamburg. At all these locations it has received not only useful criticism but in each country it has gathered round it a group of wonderful people offering serious support for the continuation of the project. They are waiting on news from Kosovo and for the green light to join the next phase of the project (the German Cultural Office has just agreed to fund the journey of the Vushtrri Red Black Box back to Kosovo). By the time this book is published the box will be back in Vushtrri – one year later than planned but with more collaborators than I could ever have hoped for.

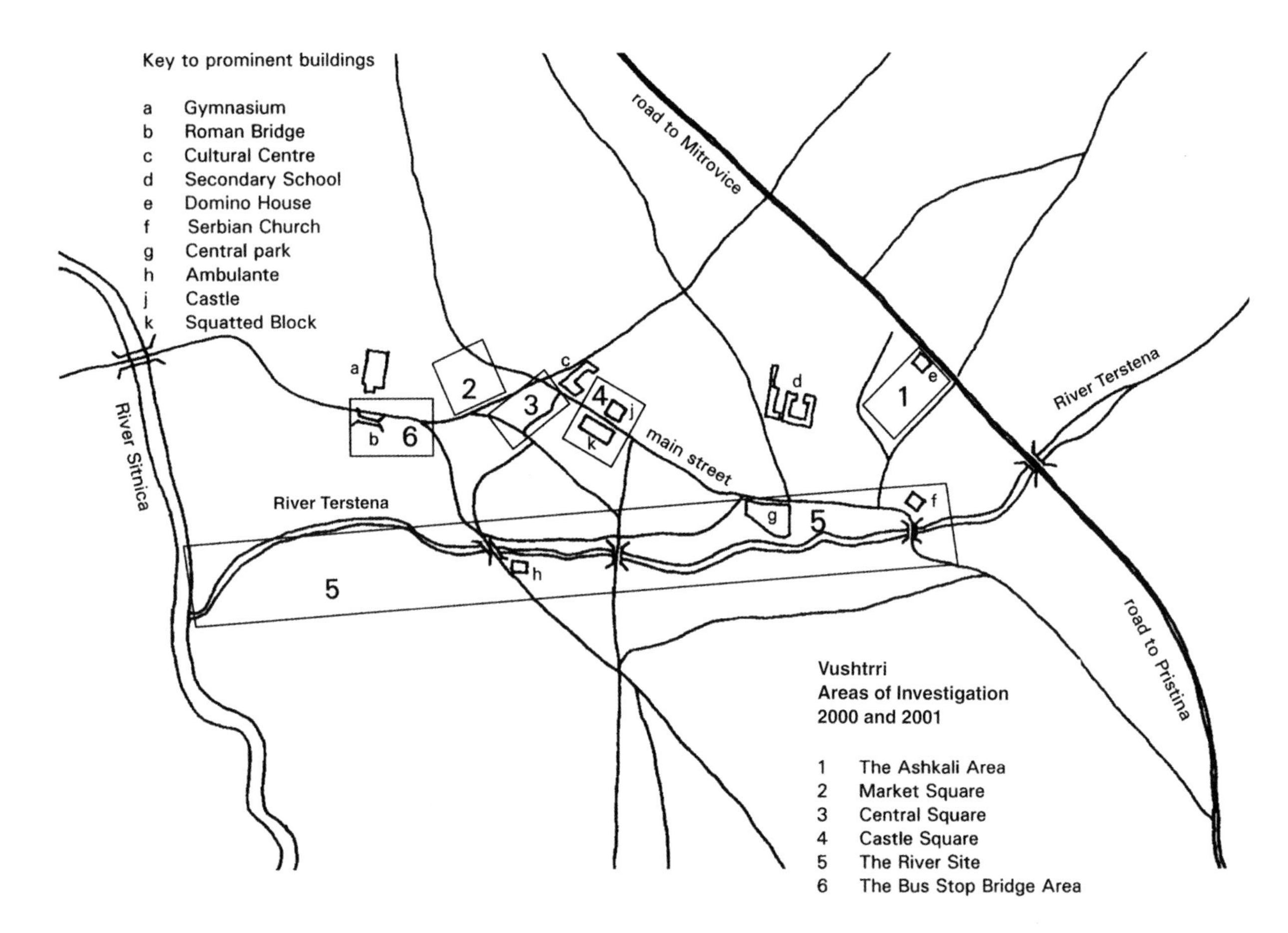

Areas of investigation in Vushtrri

Chapter 4
Student work

Vushtrri

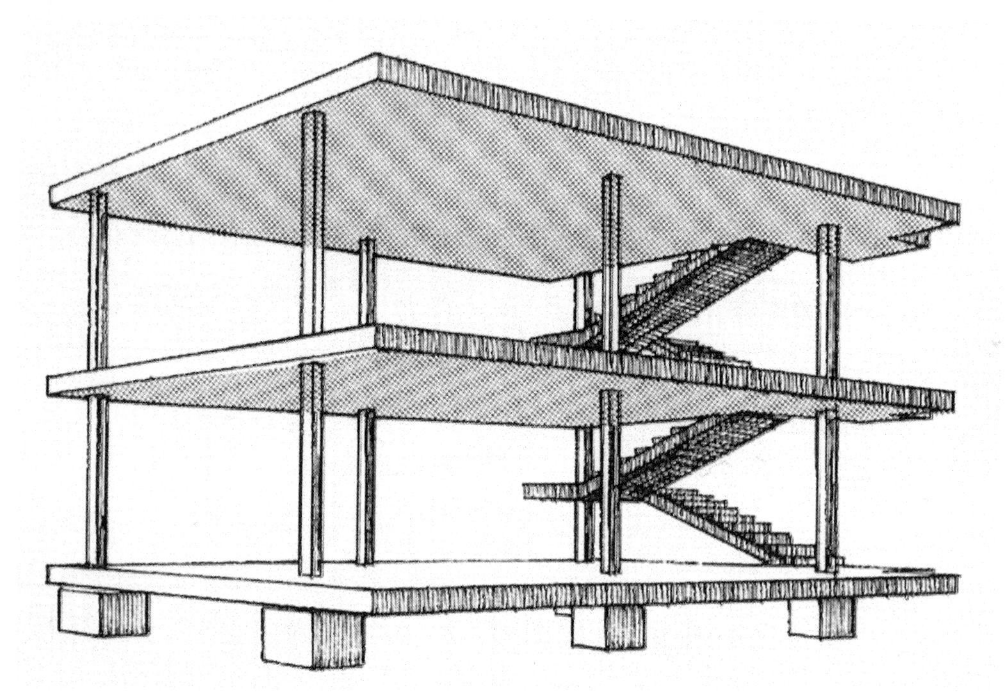

Le Corbusier's 'DOMINO' *project*

4.1 The Ashkali Area

The population of the Ashkali area was driven out during the war and their buildings systematically looted. Doors, windows, bricks and blocks were wrenched away causing structural collapse, leaving a landscape of shattered concrete and twisted reinforcement. At the end of hostilities, with the Ashkalis frightened to return, the area became a ghost town populated by the occasional squatter family. Forbidding skeletal structures remain, covered by roosting black starlings standing stark against the bright blue sky.

Within this area of fractured infrastructure and broken lives the following students chose to investigate an area they called the Gap, which surrounds a long, narrow grass and mud 'common' running from the main Pristina/Mitrovica road, on the periphery of Vushtrri, towards its centre. A few of the badly damaged houses surrounding the Gap had been reoccupied by poor rural families who had drifted into the town looking for work and camped within the ruins of the abandoned houses. Towering over them all was a gaunt, empty, ex-Serb, four-storey skeleton that the students named the Domino house. It was named after Le Corbusier's generic project, which has a similar evocative skeleton of solid floors, skinny columns and exposed stairs. At the opposite end of the Gap, away from the main road, was the largest secondary school in Vushtrri. Footpaths around the school led to Vushtrri's high street.

In November 2000 the Gap was being used for pasturing a cow, growing a few vegetables, drying washing and children's play. Water still dribbled out from burst pipes on to rubble picked over by scavengers looking to collect reusable bricks. Buildings in danger of collapse were cordoned off, but little bundles of clothes, books and other ephemera left by families fleeing for their lives remained to evoke memories of the terrible things that had happened here.

Domino from the Gap

There were several Gap-like 'commons' set back from the main road and design ideas generated for the Domino Gap could be prototypes for others. Common to all the projects was the tension between a future development taking advantage of the valuable plots alongside the fast moving and well-connected main road and the pattern of low-density, less valuable suburban streets and plots focused on Gap spaces away from the main road. Each student scheme is an alternative design idea developed from explorations on-site: both the physical survey and the students' interactions with the inhabitants.

Wrecked Ashkali house

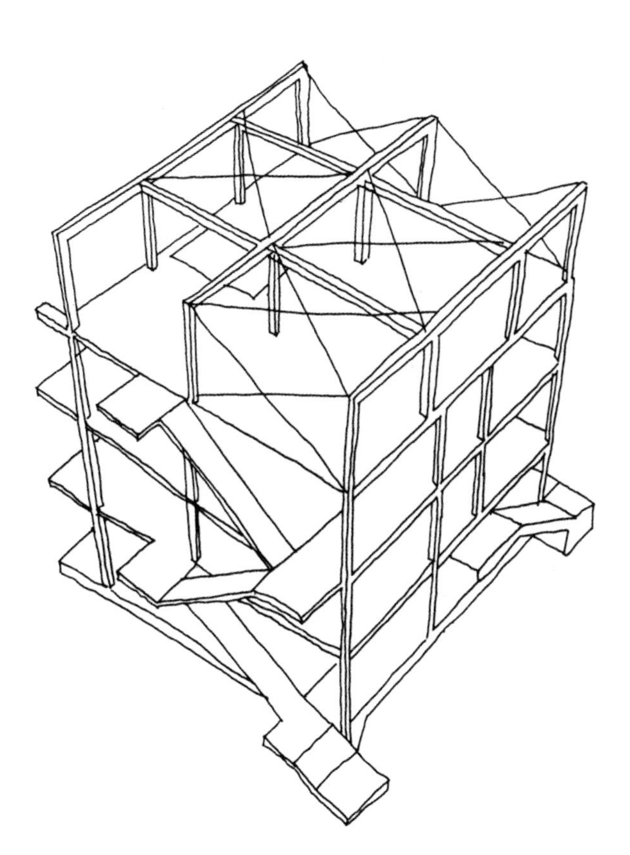

Sketch of Domino frame with external stairs added

Vera Hale

The family that had welcomed Vera and Liz into the Gap community had no address, so no future correspondence was possible, and photographs taken during the visit could not be sent back. The family's contact with a son who was a refugee in Germany was patchy and relied on messages passed on by returnees. Provoked by her study of this situation Vera proposed an idea for installing a post office in the Domino skeleton.

Post offices provide a conduit to the outside world and a social space for gossip and minor purchases. Post and telecommunications, access to welfare and other state services are provided in a relaxed open public arena alongside other high street activities. Acting as a beacon at the edge of the town, alongside the main road, the Domino's location facilitates the collection and delivery of mail.

At the base of the newly wrapped frame of the four-storey Domino, a wall of gabions (recycled concrete fragments tipped into wire cages), into which P.O. boxes are inserted, marks the side of a broad social space. Here letter writing, form filling, queuing and chance meetings can take place in safe, easy and informal spaces alongside the street.

Domino as post office transforming main road frontage

However, the occupation may be temporary. The Serb owner may return and claim his property. Long term requirements may necessitate a move elsewhere. In the short term, use as a public facility rather than a private dwelling reduces the risk of personal animosity between returnees and squatters. For all these reasons the scheme is conceived of as a lightweight, 'low tech' and short life wrapping of the existing reinforced concrete structure.

A timber frame supports the wrapping. This new frame hangs from three one-storey high timber portal frames, braced with steel rods, sitting on the existing upper floor. The outer façade is made up of recycled timber slats and the inner façade of polycarbonate sheeting. At night the building becomes a luminous box on the edge of the Gap and alongside the main road.

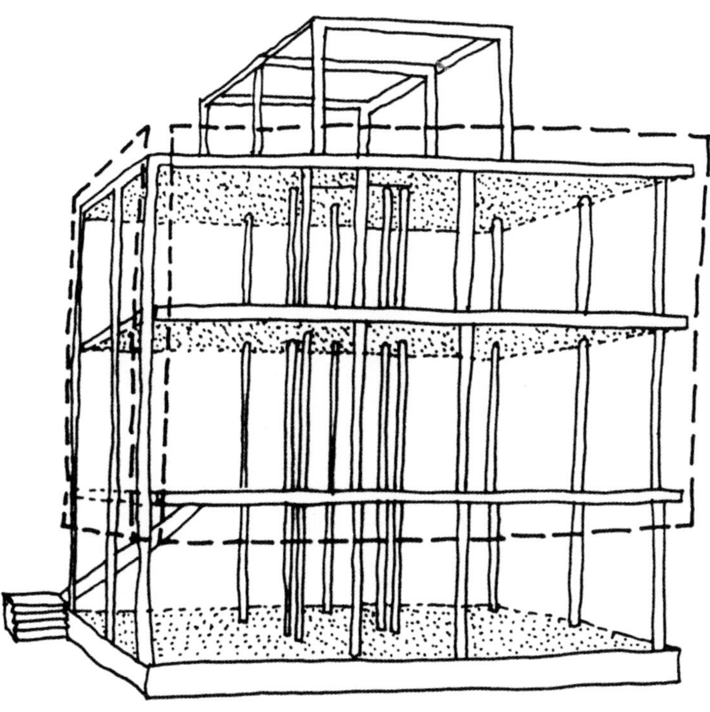

Top: Recladding the Domino frame
Bottom: Model of post office interior

Liz Crisp

Vera and Liz befriended a family squatting in a patched up, single-storey house adjacent to the base of the Domino frame. With the help of Pristina students and using gestures, and thanks to the enthusiasm of the local children facilitating communication, they recorded the daily rituals of family life, the use of the dwelling and the space immediately outside. The porch, under a roof and raised off the mud, allowed display, openness and communication to outsiders with an underlying sense of security. The kitchen, within the safety of the home, was a place where the mother passed on cooking skills to her children. In the living room young children were cared for, given confidence and encouragement and taught how to communicate with others. Outside the space was personalised by physical occupation. Piles of cleaned bricks marked the perimeter watched over by the father in his chair. Drying washing was hung up and moved around so as to stay in the sun.

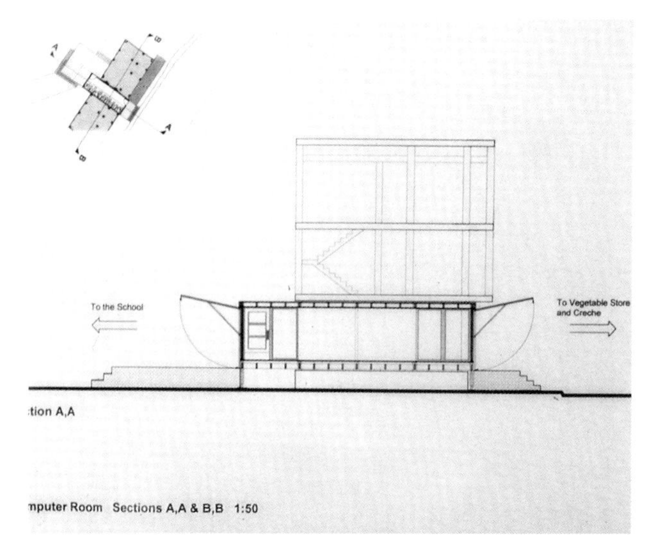

Top: Section through internet café
Bottom: Collage of Liz's internet café inserted into a looted building

Liz's proposals aimed to link the isolated Gap community to the rest of Vushtrri both culturally and spatially and she involved local school children in the project as a means of doing this. She used her observations of the squatter family as a metaphor for the use of space for education, display and communication in a secure environment. Liz's proposals facilitated the loose (low density, minimal impact, temporary) occupation of an empty, framed building with an internet/computer room; reoccupied another undamaged building as a crèche, and built a new vegetable store in the traditional manner (stone walls in mud mortar with a clay tiled, timber framed roof) to serve an array of intensively farmed allotments slotted in amongst the ruins.

A range of low key technologies were proposed. The computers and internet links were housed in a new timber box placed loosely into a gaunt, residual, patched-up reinforced concrete shell, which acted to protect and support its freshly made insertion. A substantial plinth/verandah, made from gabions filled with reused concrete, formed a platform at the entrance to the computer room encouraging conviviality, discourse and display on the doorstep.

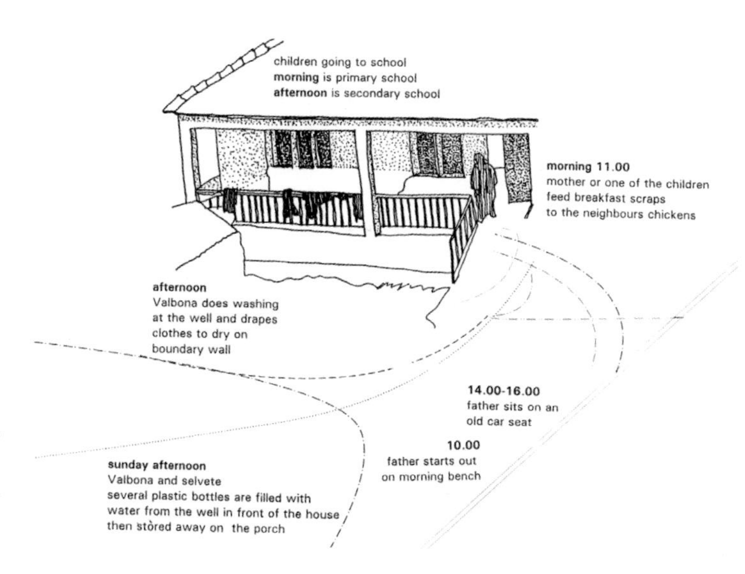

Daily movements outside Baba's house

The porch metaphor

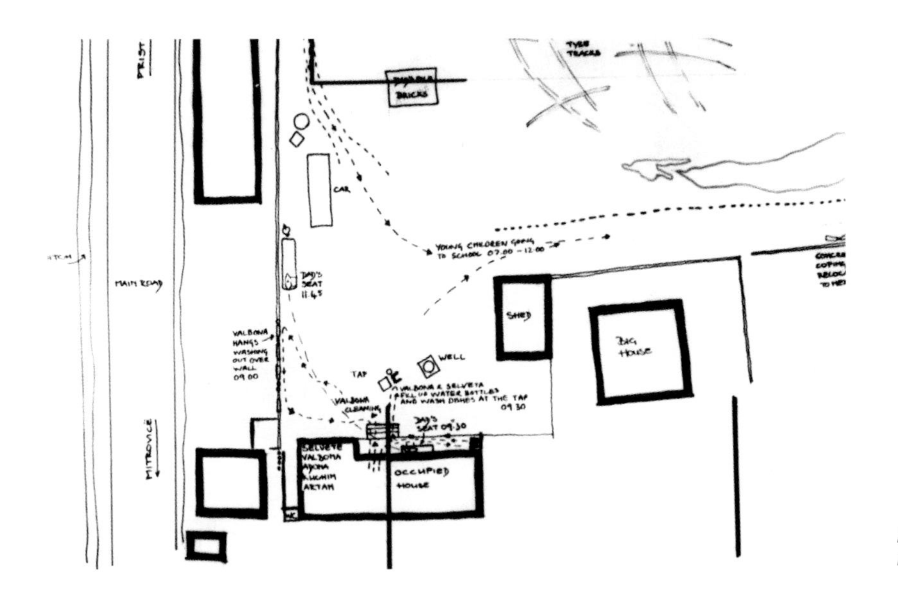

Plan of a corner of the Gap occupied by Baba and his family

Top: The crèche
Bottom: Allotments slotted in amongst the ruins

James Ross

It was James who noticed that there were several Gaps, fingers of green common pointing from the main road towards Vushtrri's main street, which could act as generating localities for groups of self sufficient communities. The Domino Gap consisted of a landscape of scattered platforms left after houses were demolished. Some off these residual platforms had been seeded to form simple allotments. James postulated a 20 year plan based on an idea about an increasingly confident community of market gardeners. The whole site was tidied up. Buildings beyond repair were cleared away – cannibalised for the benefit of those capable of rehabilitation. Bricks were cleaned, concrete was crushed and used for gravel and gabions. The resulting empty plots were fenced off to protect them from animals and then planted intensively. A vegetable market was established first along the main road and then weekly in the Gap.

Top: Collage showing the Domino frame being used as an advertising hoarding during phase 1 of the 20 year plan
Bottom: Wire frame drawing of James's Gap community

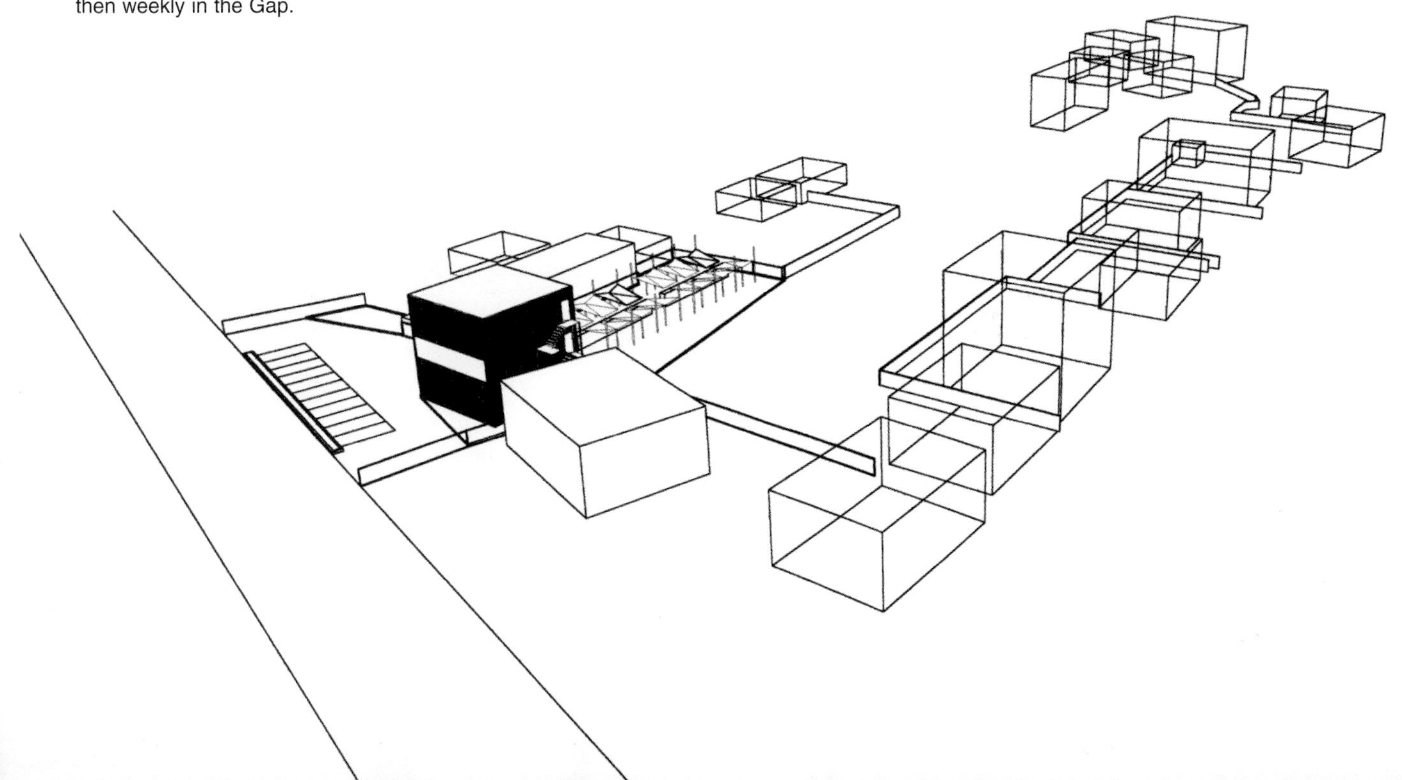

Top: Playing chess in the early market gardens
Bottom: Playing chess in the vegetable store

In among the allotments he found both children and adults playing chess. Chess is a popular national game, encouraged by the Local Authority, for which the Kosovans have international ambitions. James's Gap community would work hard and play hard. The chess game featured in a number of renderings of moments in James's 20 year development plan.

The Domino House was first to be used to advertise a weekly market, sited at its base, to people passing along the main road. Later, wrapped with polythene sheeting and timber slats, it was also to became a vegetable store with a mezzanine floor doubling as a viewing platform for chess games held below. At night, glowing in the dark, the wrapped framework became a place of entertainment, an informal mini-theatre, with chess players overlooked by an audience perched on sacks of potatoes.

Model view of the vegetable store at night

Steve Citrone

Steve discovered a continuity of construction methods running from traditional Kosovan vernacular to the FERT, beam and clay pot, floors found in the Domino House. He researched the details of current building practice and developed a project based on the celebration of existing skills, use of locally available materials and upgrading the quality of construction and performance of the various building elements. He chose to repair and infill space around a group of damaged buildings in the Gap area alongside the 'common' stretching from the Domino building back towards the centre of Vushtrri.

Poles used as scaffolding in Vushtrri

On-site exercises

Adjacent to Steve's accommodation and without boots, safety hats or fluorescent jackets, builders were constructing a suspended floor using the FERT system. After exchanging a series of furtive glances, but without a common spoken language, Steve communicated with the builders using gestures and sketches and was permitted to record the progress of their work. Beams of reinforced concrete set in terracotta channels were cast on the ground and lifted into place before extruded terracotta pots were fitted to span between them. A mesh-reinforced concrete topping bound the assembly together to form the floor. This heavy floor was cast without recourse to scarce sawn timber for formwork. The system was carefully standardised to reduce the need for engineering input. The number of reinforcing rods in the beam determined its spanning capability in metres.

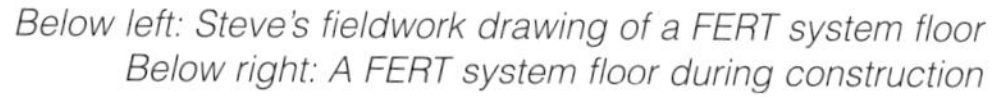

Below left: Steve's fieldwork drawing of a FERT system floor
Below right: A FERT system floor during construction

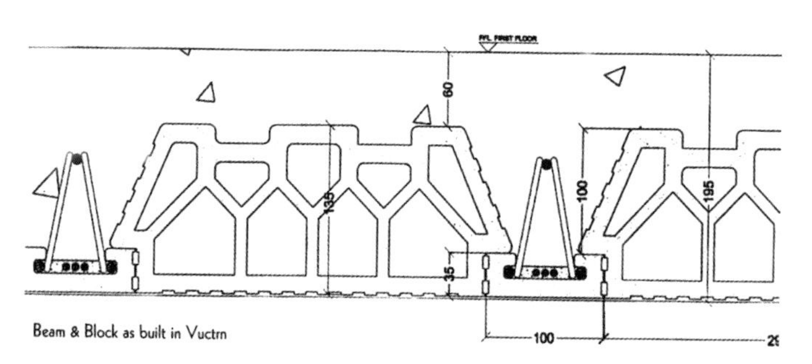

Vushtrri College of Building

The developed site proposal involved the phased construction of a school for the building trades, particularly the wet trades of mason and plasterer, where pupils would be taught existing methods and experiment with new and improved techniques.

The badly damaged single-storey houses were to be demolished and their materials recycled to repair the higher, more prominent skeletal concrete structures. The College of Building was to be sited perpendicular to the main road in a linear array of new single-storey concrete portal frames; bracketed, punctuated and serviced by rehabilitated multi-storey structures. This low roof deck became a new linear public boulevard with views over the experimental yard alongside to the re-emergent Ashkali area beyond.

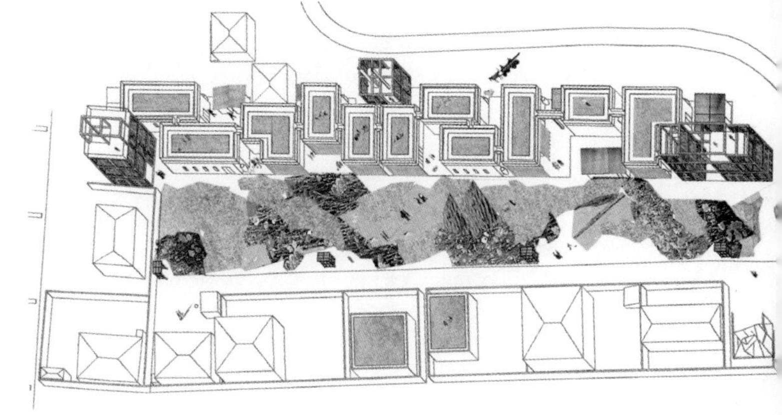

Top: The roof of Steve's building college forms a public route into town
Above: Axonometric of Steve's building college

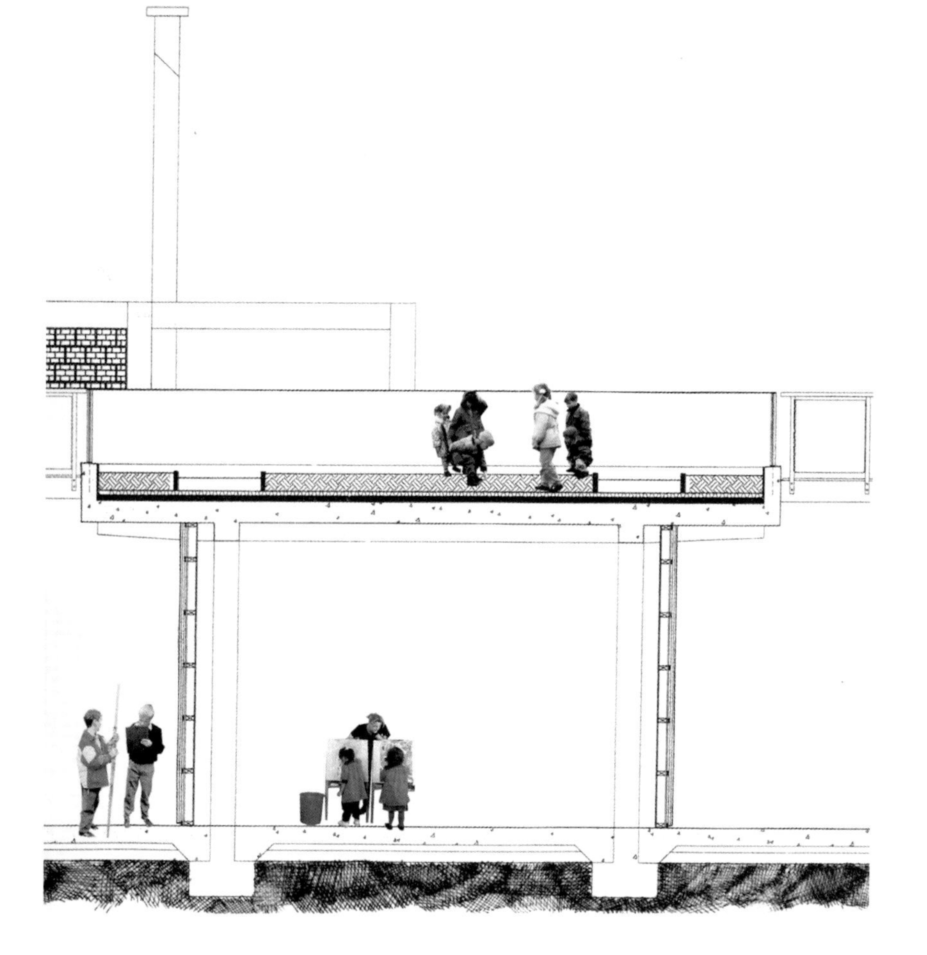

Top: Section through part of Steve's building college
Bottom: Contemporary Kosovan on-site method of decorative plasterwork

Technology

Steve proposed a number of upgrades to current and vernacular technologies to improve thermal performance. Traditional ground floor walls (of earth mixed with straw and packed between timber frames, covered on the upper floor with timber slats) were to be replaced with insulated breathing walls. Heavy grassed earth roofs were to be combined with the FERT system to improve thermal mass under the 'college roof boulevard'. Metal connectors and timber preservation would extend the remit of local timber pole technology. Extruded terracotta blocks were to be laid on edge to give ventilation with privacy in transitional areas. A local decorative technique for introducing colour and surface texture into plaster was to be encouraged.

Yenny Gunawan and Phaidon Perrakis

In 2002, following on from the initial North London University design studio in Kosovo, the author developed a design brief for students he tutored as part of the Masters Course in Vernacular Architecture of the World at Oxford Brookes University. The students were asked to rehabilitate the Domino house as a post office. Once more, design work was enriched by experimental construction work at the Centre for Alternative Technology. The project illustrated here is one of a number of examples of the Oxford Brookes students' response to this brief.

More than a post office, Yenny and Phaidon imagined the Domino building as a vehicle through which they would aim to provoke social interaction, particularly among the young people. Rather than harbouring mechanistic notions of 'fixing' the shattered building, they aimed to use the process of construction to help create a dialogue between the local people and about the building. Young people were to be involved in both the concept and practice of renovation. Involvement with the way the building would look would allow the younger generation to express their new positive aspirations for a resurgent civil society. Involvement with the construction process itself would promote familiarity and a sense of ownership of the new post office.

Top: Domino frame before cladding
Bottom: Partially clad Domino post office

Domino frame, partially clad on upper floors, with celebrating children

Left: Fully clad Domino frame with planted roof-top and celebrating children
Below: Alternative cladding systems using either vertical or horizontal secondary frames

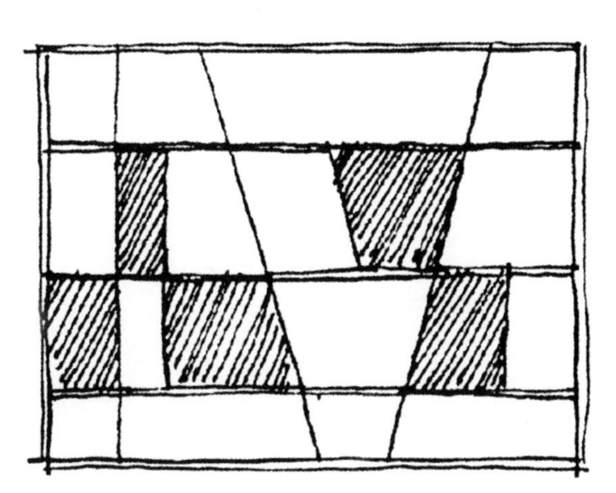

Horizontal secondary frame

Vertical secondary frame

In the Welsh workshop the students developed a temporary cladding system, based on polythene and scrap timber, which they proposed should be used as the basis for an annual, public children's workshop to reclad the Domino house. Supported by the United Nations this event would be provided with basic tools and available materials.

The building was to be rehabilitated to provide communal facilities such as post office, crèche, library and a convivial roof space. At first the children's workshop would provide a temporary skin to the building. This skin would be added to and refined with the help of succeeding workshops as appropriate. Ideas about daylighting, ventilation, view and decoration would be incorporated into the brief for each annual workshop, which in turn would inform the succeeding event.

The students produced a series of drawings illustrating their new building during the process of construction and when inhabited by day and by night.

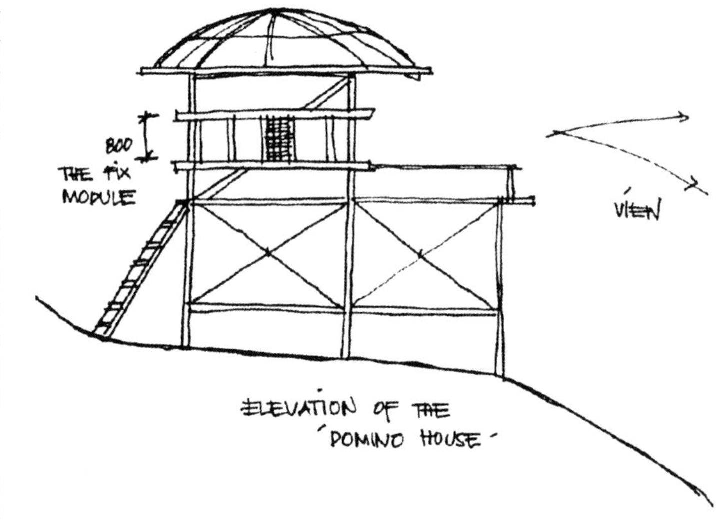

Sketch of large scale mock up of Domino frame constructed during the Wales workshop

4.2 Market Square

A building in three fragments for women

Shirin Homann-Saadat

Each Friday, Vushtrri's Market Square comes alive: it is market day. The Market Square is one of the few patches in town that has an intact asphalt ground. Due to its enclosure by adjacent buildings it has a courtyard character and a semi-private feel to it. Yet, Vushtrri's Market is underused and at night it is deserted. This might be because the square is mainly enclosed by the rear elevations of surrounding buildings, turning it into a hidden urban pocket.

The Market sits at the centre of Vushtrri, in close proximity to the town's Bus Stop and the Central Square where the main mosque used to stand. It has a large potential for public activities that need some form of spatial protection, rather than total exposure to urban life, and it could become the town's only public square to have a unique feeling of privacy.

One of its architectural boundaries is Vushtrri's impressive old hammam, a three-dimensional trace of the area's historic treasures and a forgotten public life. The hammam is a reminder of the healing qualities of water in a town that – due to the war – still suffers from the lack of a decent water supply. Currently, the hammam is rather neglected: rented by a local salesman, to store potatoes, it is waiting to be restored and revived as a piece of shared communal identity. On the market's south-western fringe is the long monolithic wall of a building that is used as a storage shed. It leads directly onto an empty patch of land at the square's north-western edge. It is this patch of land and the neighbouring 'storage wall' that I chose to be the site for my project: a building in three fragments for the women in Vushtrri.

Top: The only day that Vushtrri's market comes alive: Friday = market day
Bottom: Vushtrri's Market Square, looking towards the site at its north-western edge

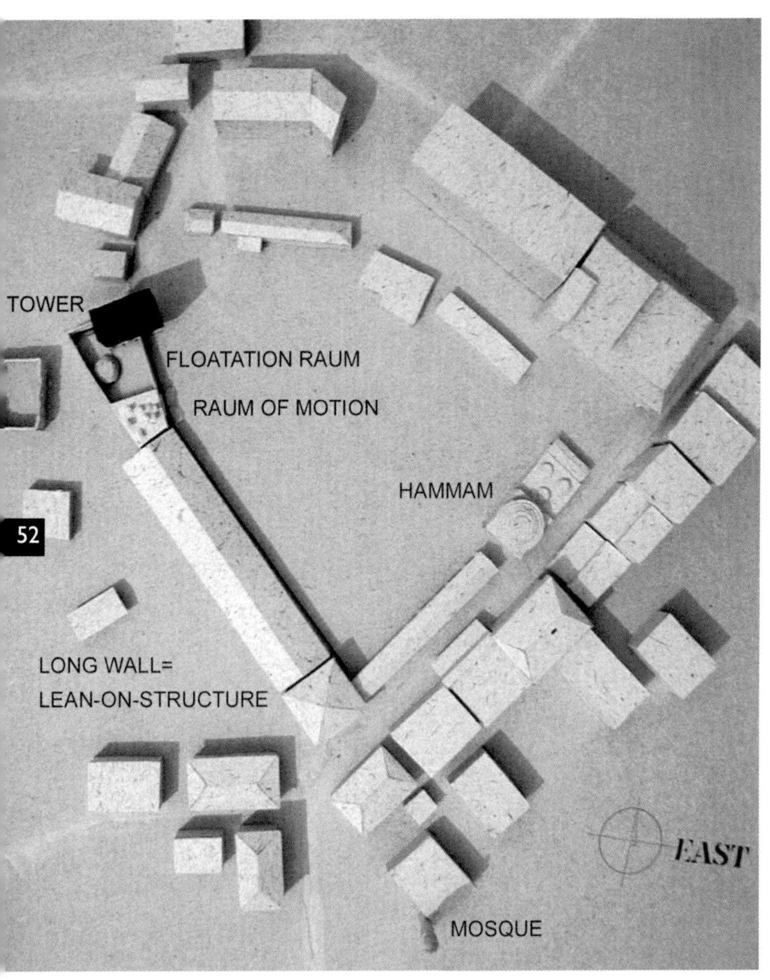

Above: 1:500 model of Vushtrri's market and the project for women (Model lives in the Vushtrri Red Black Box)
Above right: Looking at Vushtrri's hammam from above

The site is interesting because of its public yet calm qualities and because of the adjacent storage building, which could be used as the basis for a lean-to structure. The fact that the site is an empty piece of land was important to me, since under no circumstances would I have chosen a site in Kosovo that would have required further destruction of existing structures.

The project's programme grew out of conversations with local women about rapes that took place during the war. It addresses rape by assuming that it is the most private form of space invasion possible: that of the human body. As a consequence the proposal provides protected space for women only, right at the edge of Vushtrri's Market, in the centre of town.The decision to address the issue of rape in the centre of town was a direct refusal of the constant attempts to banish critical architectural programmes to the outskirts of towns, of stigmatising problematic subjects with the help of urban planning. Proposing a building for women right in the middle of Vushtrri embeds one of the most stigmatised subjects –

rape – in the centre of everyday life, provides shelter and refuge for women at the very place where they go about their daily routines. The location also offers an opportunity to hide this desire for a space of one's own: if the women felt they needed to they could simply claim to be going to the market and pop into the women's space instead.

Designed to be self-built, step-by-step or in one go, the building sequence deals with the body and with opportunities for education and work. A rammed earth perimeter wall surrounds and protects the proposal, continuing the existing storage shed wall in order to blend into its context. The proposal's landmark feature – a small tower – forms the entrance to the scheme's inner courtyard. It would house the local sewing workshop, a room of books on the top floor and space for other female businesses, to be chosen by the women in Vushtrri, on the second floor. The tower is treated with bituminous paint and has a poppy field on its roof garden, overlooking the town.

Above: Tower entrance: still from the film 'Dance your shame away'
Below left: Construction Manual, sketches, Step 4: Tidying up as a way of construction (recycling left-over soil for rammed earth walls)
Below right: Sewing workshop in Vushtrri

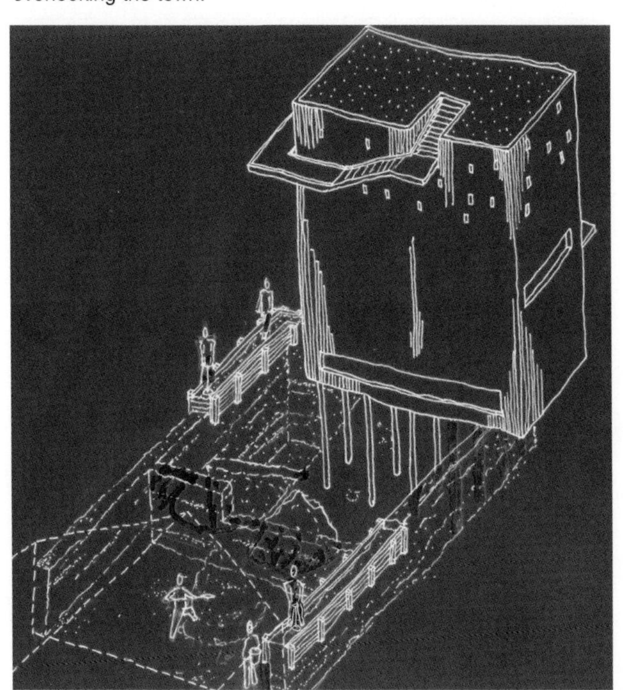

Painting of the Room of Motion for the film 'Dance your shame away'

The scheme's second fragment is a flotation space, a room-sized echo of the hammam on the other side of the market. It is sunk into the ground for climatic reasons, and for sound insulation, and offers space in which women are not exposed to the usual cacophonies. In the flotation room they can keep their sense perception to a minimum. Originally developed to cure back pains and nervous disorders, the salt-water mix in the flotation tanks has soothing and healing effects on stress-related disorders. The experience of floating, i.e. not being aware of one's body weight or surrounding temperature, might be one alternative to verbal attempts to deal with the issue of rape. A frequent problem caused by rape is the silence that follows. An inability to talk about the crime should be respected: concentrating on regaining a feeling for one's body might be helpful as a first attempt to address the after-effects of rape.

On the way down into the flotation room one enters a small sunken courtyard full of herbs, a scent-threshold between the everyday and the float. The proposal's third room represents the opposite of the flotation room. The room of motion is a lean-to structure, to the existing storage shed to its south, and provides space for all sorts of movements and sounds. It is the only fragment that leads directly back onto the market and it provides a space where women can let all (e)motions out. Constructed of rammed earth it has particularly thick walls towards the courtyard of the proposal so that the inner peace of the building sequence is not disturbed. The room of motion houses a small punch bag forest and can be opened up in such a way that the women would be able to use it as a performance space that would melt back into the town's market.

While the tower's educational and work orientated programme is a physical way of saying 'let's look ahead', the two other building fragments deal directly with the body, though in opposite ways. Read as one piece the three building fragments try to be sheltering islands for the women of Vushtrri, who were not 'innocent civilians' during the war but who were actual battlefields.The whole project tries to work against the most private space invasion/rape by making space for the women only.

An ambition like this is perhaps bound to fail. But if only a few women were to start refusing the term 'victim' and turn the experience of violence, in just one of many pointless wars, into a new means of looking ahead the project will not have been totally in vain.

Painting of a floating girl for the film 'Dance your shame away'

4.3 Reclaiming the streets: Central Square

During the dog days of thuggery, small bands of gunmen had roamed the streets shooting at will. The town's inhabitants hid behind high compound walls, posted look outs to peer down from the rooftops and monitored the progress of the gunmen. If they tried to enter, to smash down a front door or climb a wall, warning was given. Holes were cut in party walls to allow escape through the warren of adjoining properties. The streets were a wasteland and the neighbourhood of contiguous compounds a tortuous maze, wherein, with a bit of luck and the help of your friends next door, you could hide from the terror gangs.

After the war, holes in the party walls were sealed up and new holes were cut in the outside walls to provide small windows through which cigarettes, CDs and sweets could be sold. The streets became once more the place of meeting and exchange but also the place of anger at what had happened: a place of demonstration.

Woo Song

Woo Song recorded the progress of a series of demonstrations, which took place in Vushtrri in October 2000, to protest at the continued detention of Albanian Kosovars by the Serbs. He showed how the character of the crowd would change to reflect the function of the space they were passing when skirting the larger public areas. The cemetery would be passed in silence; fathers would call to their children in the park; the football ground would evoke a lighter mood with less bitter memories. The crowd moved up and down the main street missing out key public spaces. Woo proposed a new route, a circuit for protest or celebration, a carnival loop, sweeping through and past the key public spaces in the town.

Processions would still start and finish in the Central Square. At the time of the Studio this space was bounded on the higher side by the Cultural Centre (a constructivist edifice whose imposing steps were used as a speaker's podium), on its lower side by the United Nations Interim Administration Offices and on its other two sloping sides by sweeping, less imposing shopping streets. In the centre there remained a jumble of offices to one side of a huge, empty hole: a bare, rubble strewn slope surrounded by a perimeter fence of flat metal sheets. This used to be the site of Vushtrri's central mosque. It had been dynamited and then demolished before being removed stone by stone by the Serbs.The site was left as a grassy patch the size of a football pitch boarded off on all sides: an empty space in the centre of Central Square.

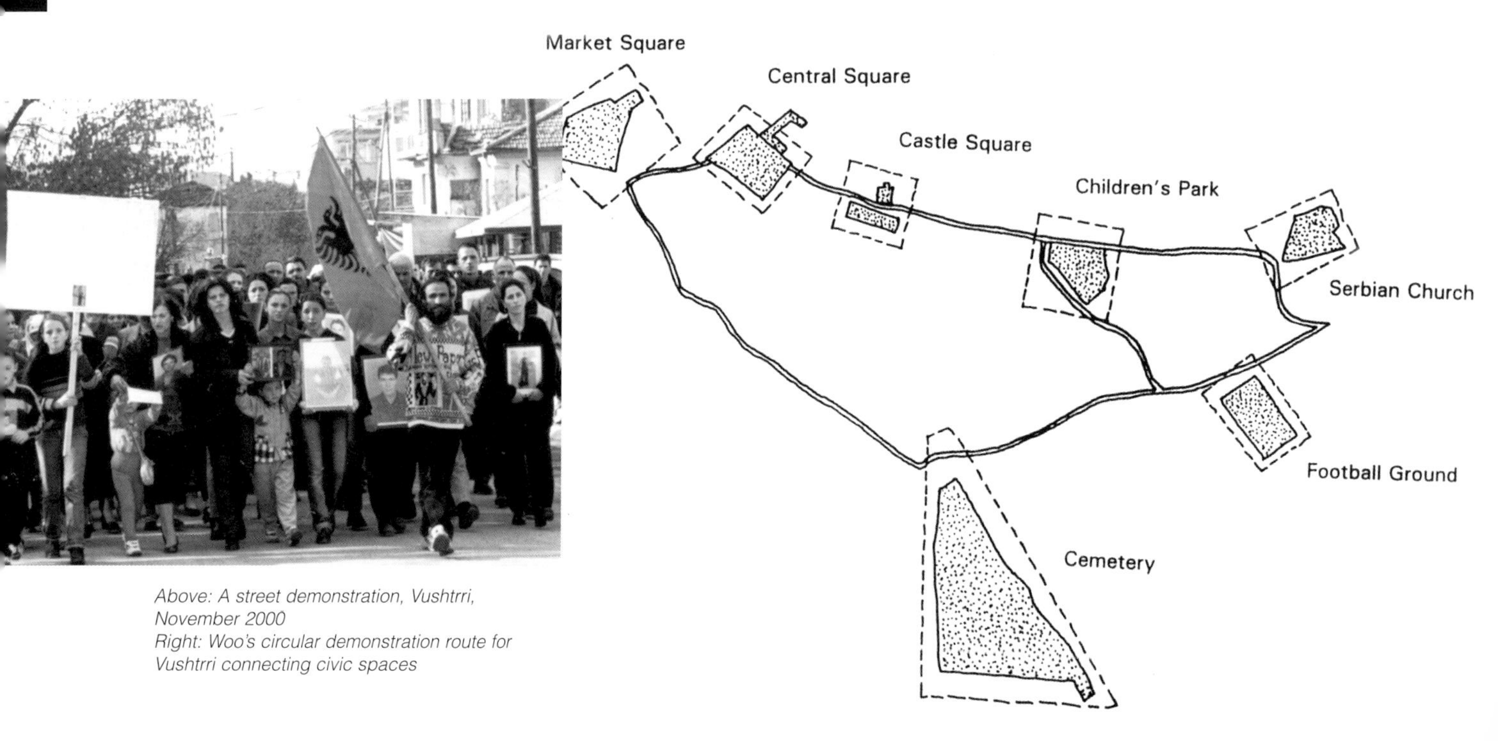

Above: A street demonstration, Vushtrri, November 2000
Right: Woo's circular demonstration route for Vushtrri connecting civic spaces

Jason Wells

Jason Wells chose the Central Square site to create a new public space. After conversations with local people, Jason decided against rebuilding the central mosque. Saudi Arabia, the local KFOR occupying force, had offered to rebuild the mosque on an even larger scale. Locals had asked them to build schools instead. The war had affirmed the population's preference for local neighbourhood mosques embedded within the townscape. The larger urban scale public buildings should be for education not religion. Jason chose to design a public library for the square.

Rather than provide an inward looking, contemplative, orthogonal, car-free area bounded by controlled facades, Jason continued the pattern of public space evident in Vushtrri: public space bounded on one or more sides by a vehicular route. The construction of the library would be phased, becoming over time a more permanent and familiar structure facing down across sloping ground alongside the busy shopping street that leads to the market.

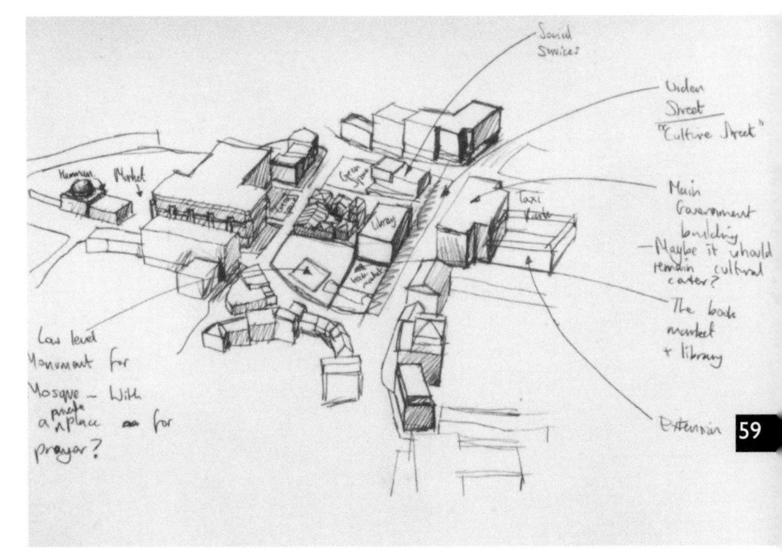

Above: Sketch showing proposal to place a mobile library base in the Central Square
Below: Vacant plot in the Central Square where Vushtrri's mosque used to stand

Above: Rendering of proposed book exchange library in Central Square
Below: Final scheme for central library building

4.4 Castle Square

Back along the main street from the Central Square one is confronted by a curious juxtaposition. Facing each other, across a jumble of road and tarmac, are two buildings which couldn't be more different. To the south, overlooking the town, rises a squatted, high-rise apartment block. Opposite, the high but crumbling outer stone wall of the 'Turkish' Castle, the oldest building in Vushtrri, faces the square from the north. The first has a sound fabric, a solid material but negative cultural value, whilst the second is a crumbling heap of ancient stones lodged in the constructed collective memory as signifying a more secure and positive past, materially bankrupt but culturally meaningful.

Front and back, inside and out

The reinforced concrete framed structure of the apartment block was completed and partially infilled before being abandoned during the war. Returning refugees occupied the building, which was still without water or mains services. Rumour has it that, for want of sanitation, the basement is filled with unhealthy detritus of all kinds. Within the private zones occupied by the returnees, however, decoration is sumptuous and cosy: a high-rise haven up above the tumult below, a jewel box on stilts. The southern façade, which perches over the town, invites an unnecessary dose of solar gain in the summer, whilst the northern face completely blocks out the sun from the public space in front of the Castle.

The Castle is a hollow, empty shell consisting of a raised, dank-walled courtyard on a slope with a squat tower on the uphill side. At the lower end, facing the public space, this edifice imposes itself onto the main street causing a kink in the flow of pedestrians and traffic. It is possible to escape through the back of the tower uphill into a maze of side streets, rubbish strewn ditches, alleys and back yards.

Raymond Leung

Ray Leung proposed a new public square between the Castle and the squatted block. He skinned the tower block at the lower levels, extending a level 'common ground' through the newly exposed columns, to open the square to a view over the rooftops of the town and to glimpses of sunlight. A terraced pedestrian route, a potential tourist narrative, was opened up – running downhill from the Castle through the newly revealed townscape to the market place. The square itself would become a play-box of ephemeral activities, servicing a new theatre/cinema within the Castle courtyard and pandering to the throngs of energetic young people which now, in the relative safety of postwar Vushtrri, feel free to promenade down the main street in the evenings and at weekends.

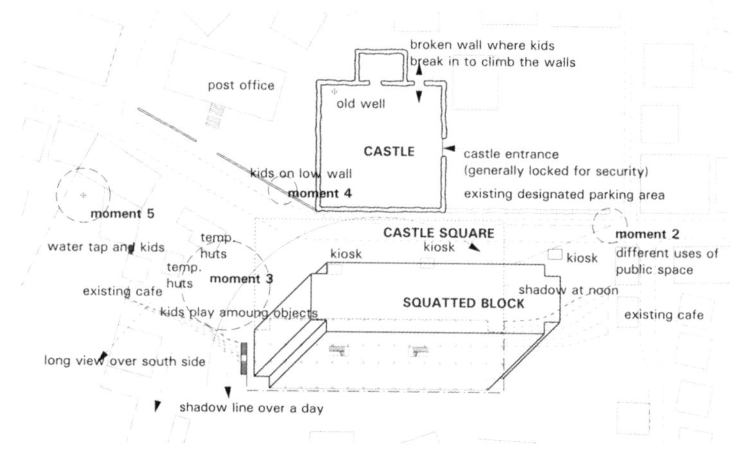

Axonometric of Castle Square showing observed activity

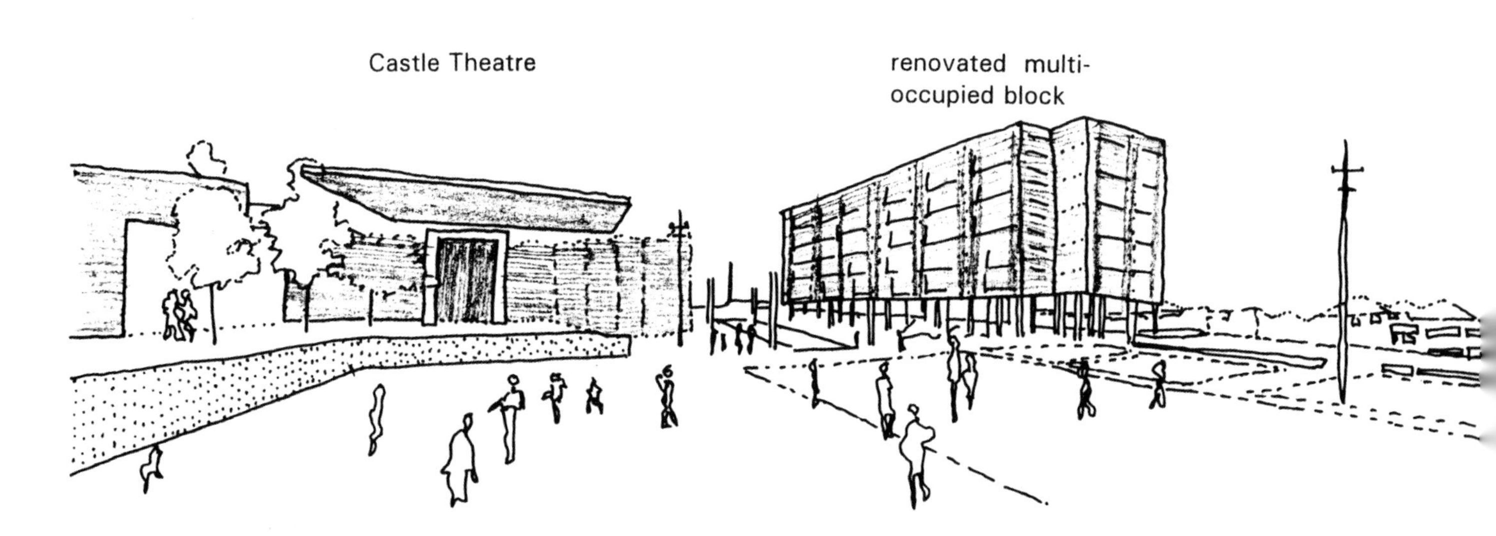

Sketch of new Castle Square and squatted block from the post office

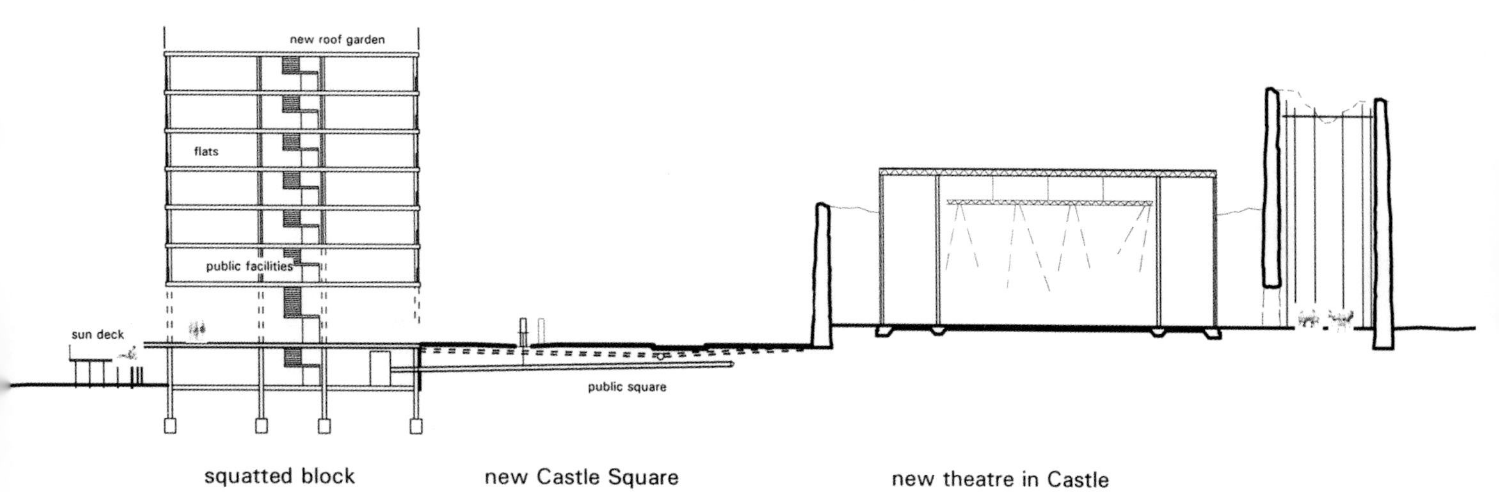

Existing informal uses for play, conversation and café life were noted and are encouraged. The surface of the square has been divided into zones, which use different paving materials to indicate their primary function. Crushed, polished concrete or fired brick were proposed for areas designed for casual gathering and more focused assembly, sand for children's play areas and timber decking for lounging in the sun.

One of the observed moments located in the corner of the new square, outside the central post office, beside the Castle walls has been developed into a 'play-box', a place where adults could congregate and children play.

Traffic access has been limited to certain hours of the day to encourage a leisurely pedestrian syncopation between sun and shade. The municipal nature of the intervention, marked by the carefully zoned paving is intended to provide an optimistic, post-conflict, neutral common ground for all comers, poised between tradition and modernity.

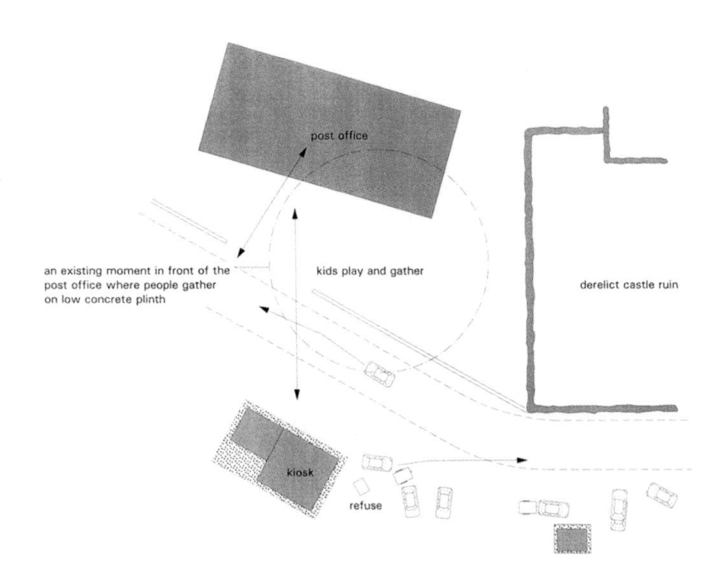

Top: Section through castle (used as theatre) new Castle Square and transformed squatted block
Bottom; Plan diagram of space in front of post office used for adult gathering and child play

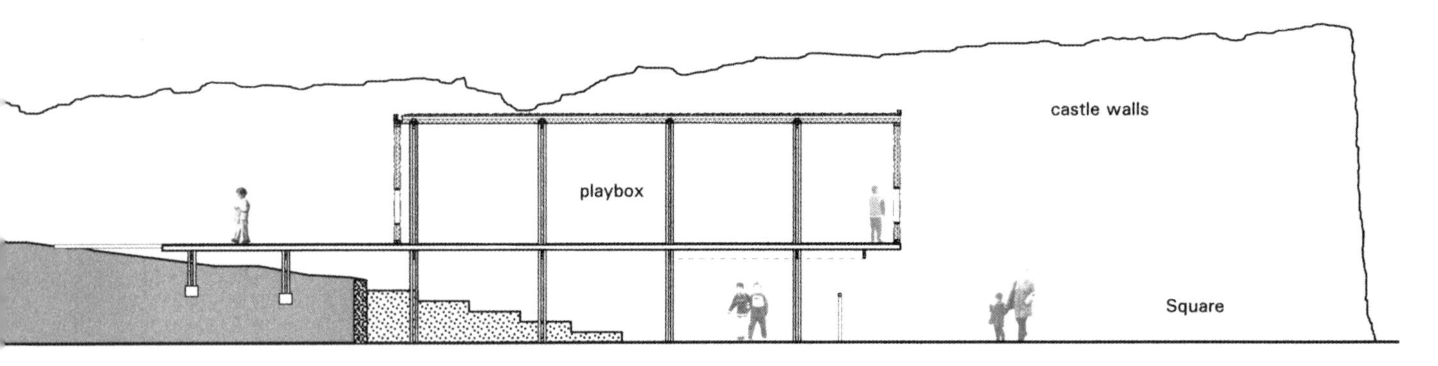

Above: Section through 'playbox' building against Castle wall
Below: Elevation of 'playbox'

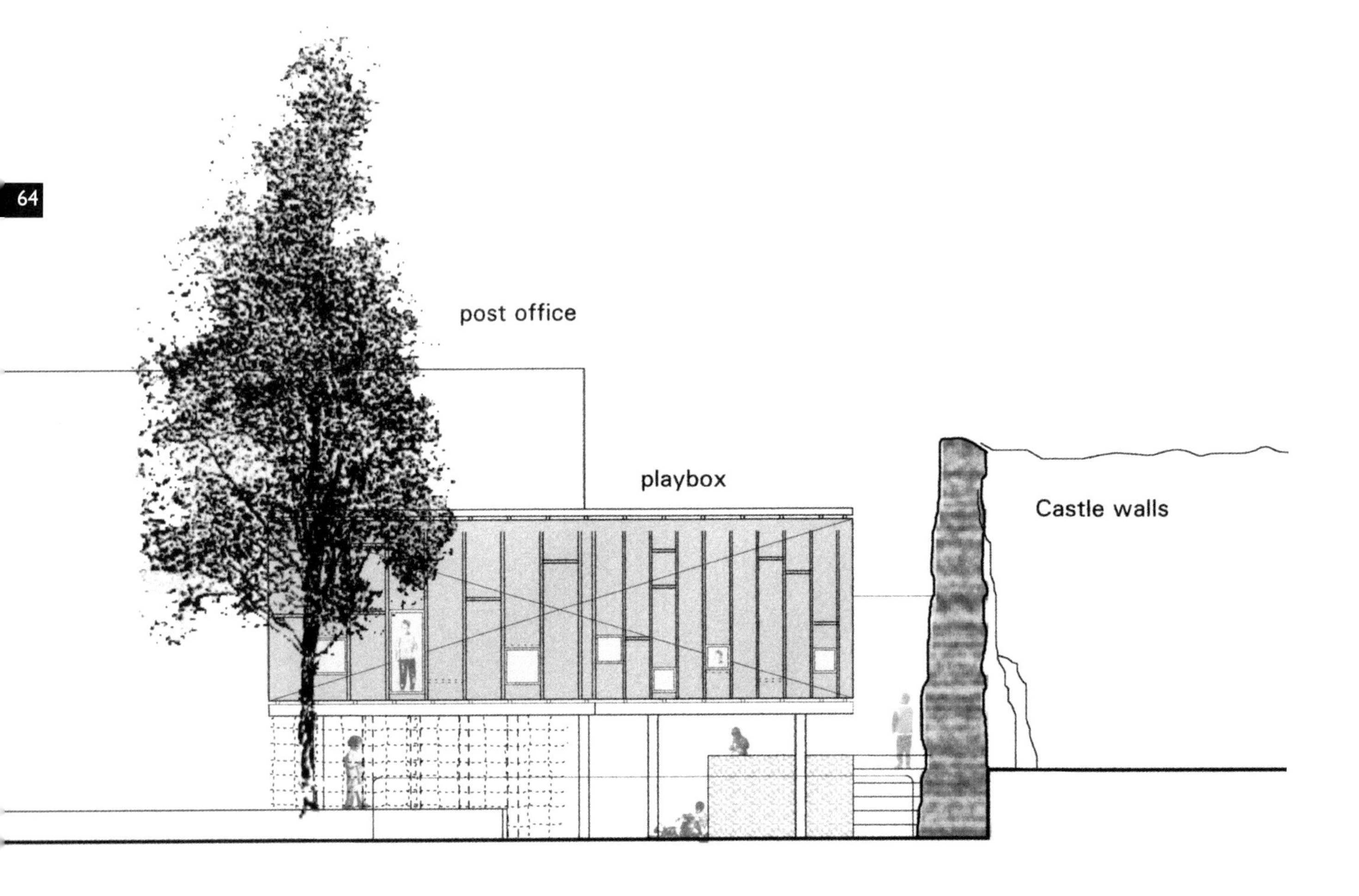

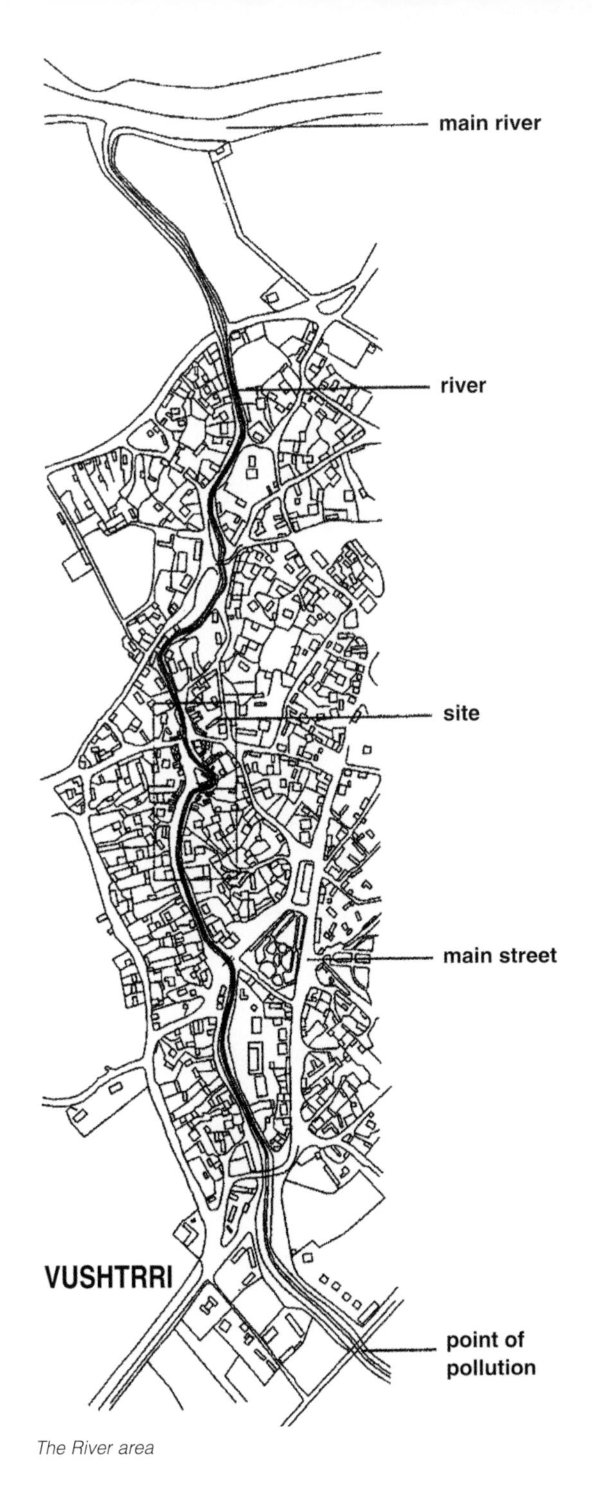

The River area

4.5 The River Site

The river Terstena cuts diagonally through the back of Vushtrri from east to west. Properties face away from it. It is a small backwater, a source of pollution and a receptacle for community refuse. In a more developed country it would have been covered over and turned into a public sewer. Children play among the rubbish and walk to school along its banks. There is land here suitable for development if the annual floods could be controlled.

The river Terstena is clean before it heads into town. As it leaves, it joins the larger and even more polluted (by the Oblic power station) river Sitnica, which flows north to join the Danube. Schemes sited along the river focused on issues of reducing pollution, providing clean water, children's safe play and providing car-free public amenity space. Most schemes envisaged damming the river upstream, channelling the water between tight banks and using reed bed technology to green the landscape and clean surface water. Students were intrigued by the starlings roosting over the trees at sunset, energised and shocked by the children's stories and both attracted and repelled by the ribbon of polluted but alluring spaces.

Andrew Fortune

On-site exercise

Andrew noticed the potential of the site on the first day of his visit. He galvanised the children of the area to collect discarded plastic bottles and within 24 hours they had constructed a bottle bridge, bound with reinforcing rods and plastic straps. The poster, the product and, above all, the process gained Andrew access to the river community via the children. He was able to discuss with them their fears and aspirations for the river area.

Back in the UK Andrew took his experiments with plastic bottles further, inventing a roof panel. Continuing his hands-on investigations, he inserted glass bottle panels into an earth brick wall reinforced against earthquake damage by timber string courses.

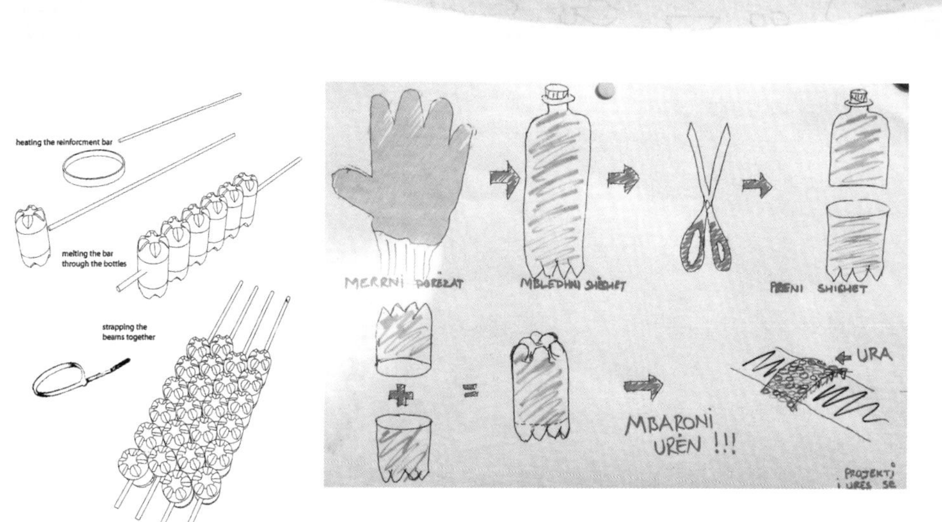

Above: Boy crossing the plastic bottle bridge
Below left: Diagram showing how to assemble the plastic bottle bridge
Below right: Poster showing how to construct the bridge

Water strategy
Having proposed damming the river to control the flood, Andrew went on to ameliorate the deficit in mains water supply. He proposed an aqueduct system, a series of ring mains, which would collect water from roofs, store it in reservoirs or in local water tanks, filter it and redistribute it to households in the neighbourhood.

New public space
At a crossing of river, road and aqueduct, Andrew proposed a new public space covering the river. Existing run-down sheds on a tight bend were converted into children's workshops. Separating the workshops from the square and providing one edge to it, the aqueduct became a two-storey, 50 metre long terrace facing a row of existing shops on the opposite edge. In the centre of the new square, clean, filtered fresh water was available from public water tanks.

Along the edge of the square the aqueduct allowed access to pipework for maintenance, but at the same time it served as a rain water storage and filtration unit, and provided the space for a cycle repair workshop (for riverbank cyclists), a post office and terrace café beneath it.

Technology
The technology involved was remarkable for its:

- hands-on characteristics (experimentation with bottles and earth bricks),
- finding new uses for waste materials (bottles),
- the use of Portland Cement as much for its water resistance as its strength (ferro-cement water tanks and sand filters),
- and its acceptance of the value of locally developed construction methods as the starting point for material proposals (FERT system).

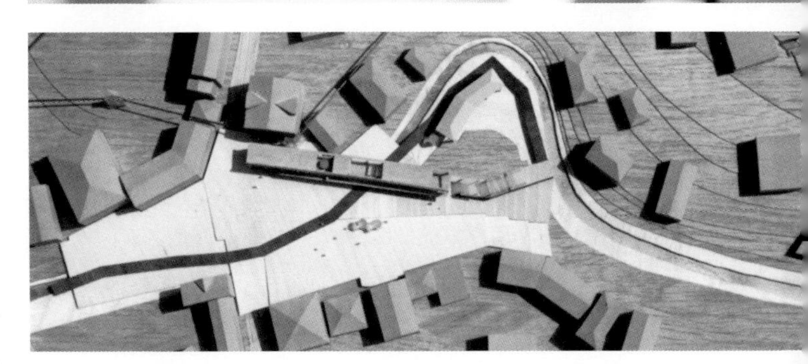

Top: Kids kitted out with gloves to work on the plastic bottle bridge
Middle: Model of new inhabited aquaduct in the square
Bottom: New square on the river

Collage of the new public square

New inhabited aquaduct on the river

Ben Brown

An idea for a roof combining a number of disparate structures

Combing the riverside for useful pockets of potential, Ben divided the ruins and vegetable plots into 'Sustainable, Locally Managed Clusters' (SLACS). On the scale of large family groupings rather than whole neighbourhoods, these collections of abandoned building shells and open spaces were to be occupied and run by returning groups of displaced people on a co-operative basis.

To test out this idea Ben chose four looted building shells standing on the riverbank at the start of a riverside walk. His organising design idea was to provide a roof canopy over-sailing and uniting this motley collection of damaged reinforced concrete frames, producing a multi-occupied, jointly managed, four-storey urban block. Extra stairs were inserted in the gaps between blocks giving direct access to the upper floor. Between the roof and the frames, an extra floor of lightweight construction provided

Ruined stairs within existing reinforced concrete frame

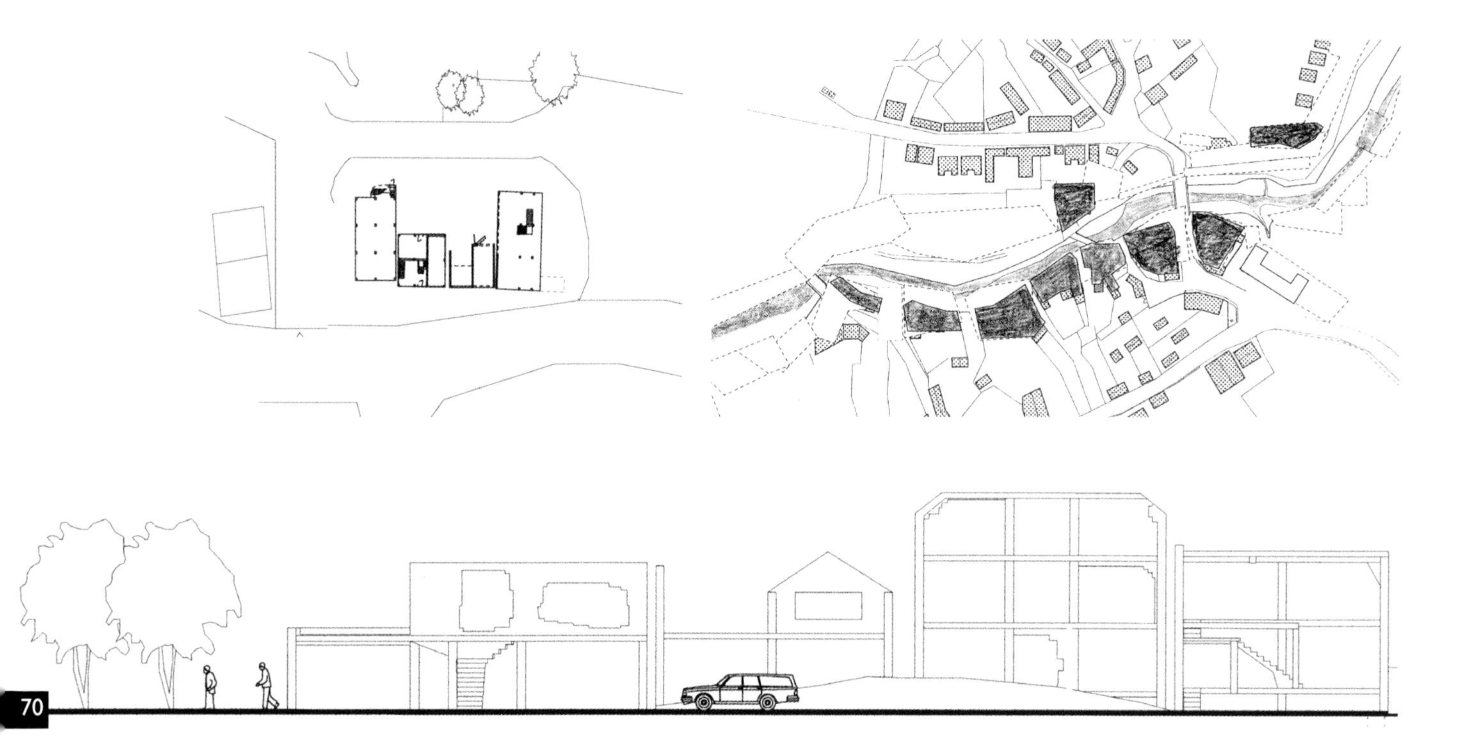

temporary accommodation for the recently arrived. Rainwater would be collected from the roof. The frames themselves were to be provided with clean water and electricity and progressively in-filled with permanently constructed shops, launderettes and cafés on the ground floor and flats above them.

In this way Ben developed a new urban prototype mixing the established FERT type construction with an upper, see-through floor of community facilities under a long span roof. The roof type familiar to FERT construction uses Marseilles interlocking tiles on a timber roof at a pitch of 30 degrees or more, which for longer spans means a very high, overbearing roof. By using 'Onduline' a corrugated tarred paper product under the tiles, Ben was able to use the local tiles at a reduced pitch of just 15 degrees making the roof low and light, as well as long-span.

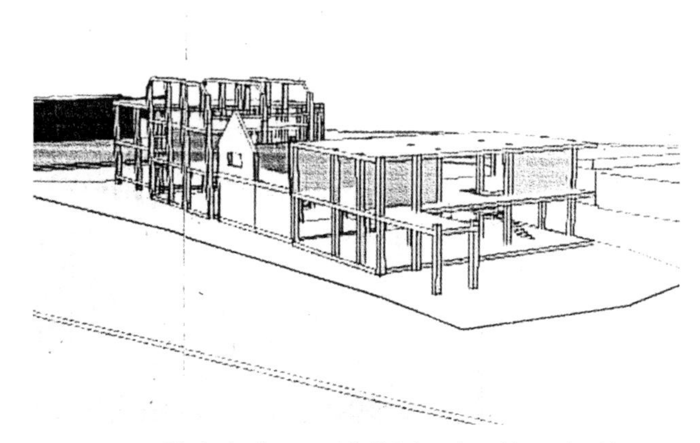

Clockwise from top left: Existing plan of four ruined houses; Sustainable self managed river communities; Existing elevation of four ruined houses; Perspective from the east of four ruined houses

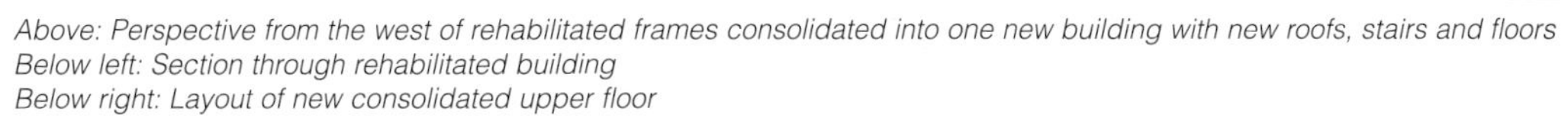

Above: Perspective from the west of rehabilitated frames consolidated into one new building with new roofs, stairs and floors
Below left: Section through rehabilitated building
Below right: Layout of new consolidated upper floor

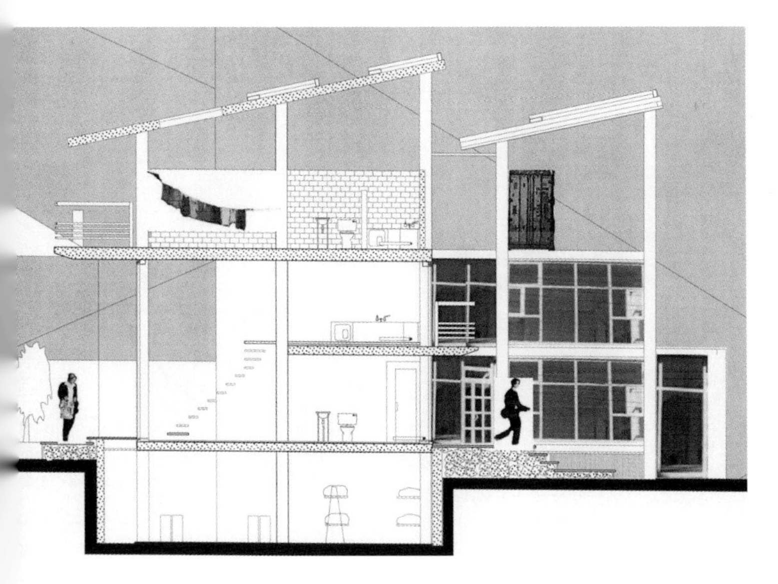

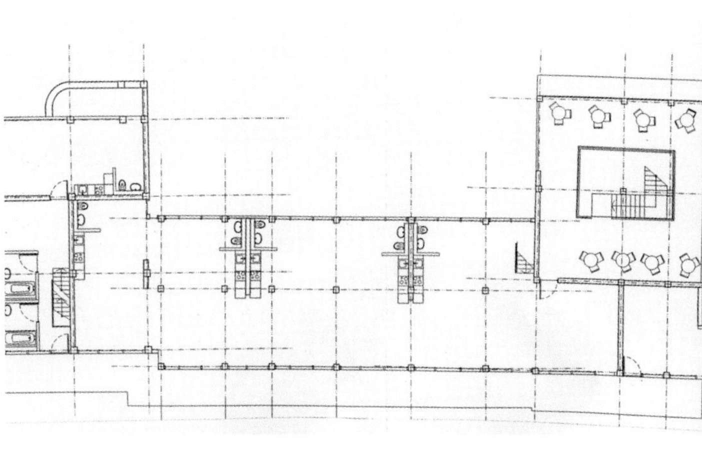

Keith Smith

For Keith, the ribbon of river was a place of quiet, away from roads and crowds. It could give access to houses but it was not a through route for cars. The further people moved away from the road crossings the easier it was for them to carry on peacefully spreading their daily activities out across the river itself, and the pathways on either side, without interference.

Keith noticed a marked difference between the perceived freedom of women in the small town of Vushtrri compared to the women of Pristina. In Vushtrri he saw women now and again behind counters in family owned groceries, newsagents and clothes shops. On the street they were shy and curious but rarely on their own. Unlike the men they shied away from having their photograph taken. He noticed that in public play spaces, boys seemed to take over the play space leaving girls on the periphery.

Sketch of river site as existing

Sketch of river site with womens' centre

Despite the difficulty of obtaining clean water, people were continually washing their front path clear of the dust and mud of the streets and then having to move rubbish away from their drain pipes to allow the muddy water to run freely into the polluted river.

Keith resolved to tackle river pollution and provide clean water for women and children on a secluded site midway between two road crossings. He proposed a timber clad, reinforced concrete framed tower at the side of the river with a 'carpet' of activities stretched out over the river to the other side. This 'Hammam Tower', a replacement for an older hammam now used as a potato store at the side of the market, would use the cleaned water to purify in a spiritual (as in Islamic ablution rituals) as well as a physical sense.

Above: Elevation of women's centre
Below: A row of Kosovan beehives

The tower consists of three square boxes, a Kosovan beehive of activities, stacked on the openwork legs of the laundry and lobby on the ground floor. Moving upwards through the baths (first floor box) and sauna/steam room (second floor box), the teaching and prayer box (top floor) is skewed to face Mecca. The roof collects water and solar energy. Out in the open, raised above the mud and dust on a 'carpet' of decking – which is stretched out over the river and its machinery for filtration, cleaning and purification – there would be a small children's pool, a fountain and space to display and sell handicrafts.

Chris Hale

Chris chose to investigate the tree-less flood plain at the junction of the Terstena and the Sitnica. This area of bunds, grassland and birds is a wasteland, apart from the town, lonely and windswept, which forms the end of the proposed new pathway and cycle route through the town. Down below this plain, out of view, at the polluted water's edge children play exposed to numerous dangers. An erstwhile fisherman complained that for a few years now he had been unable to find any fish: the effluent from the Oblic power station had killed them all off. Lulezim, a young teenager, told Chris a story, using sign language, drawings and then through translation. He related how his friend had drowned the previous week, embroidering the tale with a description of how the police car had crashed in flames rushing to save him. The police confirmed that six children had drowned at this spot within the last year.

On an urban scale, Chris envisaged a patchwork quilt of reed beds (to treat the effluent from a sewerage system proposed for the whole of the Terstena River Site) interspersed with allotments – exploiting the compost produced to provide vegetables for Vushtrri. This was to be combined with safe water-based leisure facilities for young people. A riverside promenade would connect the market gardens with a swimming pool and supervised beach area, itself protected from pollution by a series of treatment and settlement ponds. Strategically this scheme might become a model for sustainable development, combining riverside reclamation, market gardening, leisure and pollution control along the river Sitnica.

Gabion walls collaged on to the flood plain of the River Sitnica with Lulezim in the foreground

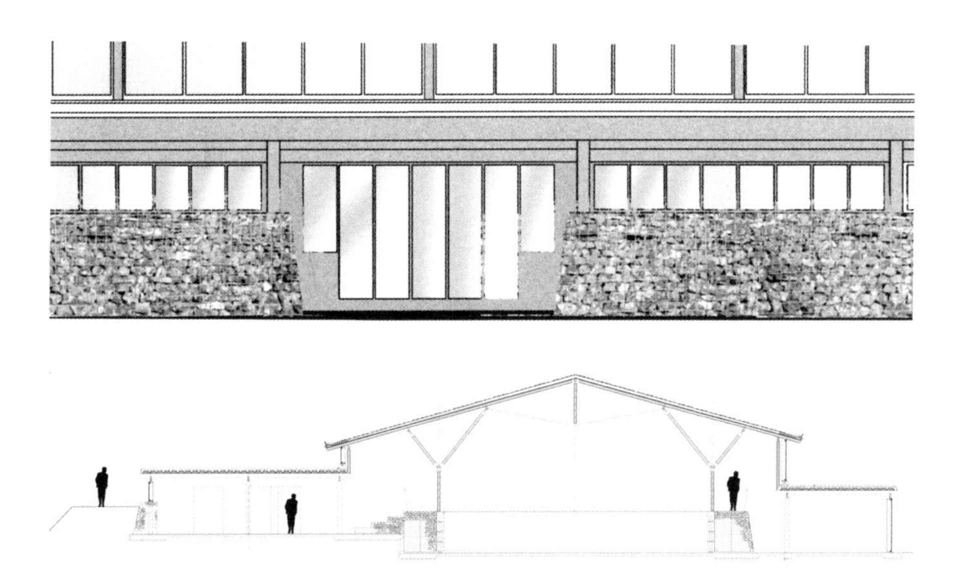

Top: Part elevation of swimming pool showing incorporation of gabion plinth walls
Middle: Section through swimming pool showing timber pole structure supporting Onduline roof amongst gabion bund walls
Bottom: Model of gabion walls made at CAT

The walls of the swimming pool would be constructed from gabions, using rubble recycled from the Ashkali Area. The long span portal roof frames were to be made from local round pole timber with metal connections. The roof covering would be partly grass to match the landscape and partly 'Onduline' backed Marseilles tiles at 15 degrees pitch as described in Ben Brown's proposal. Solar panels would heat the pool water.

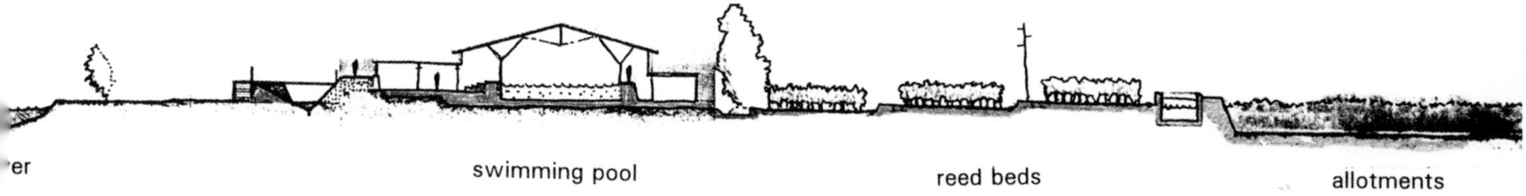

er swimming pool reed beds allotments

Section through allotments, reed beds, swimming pool and flood plain

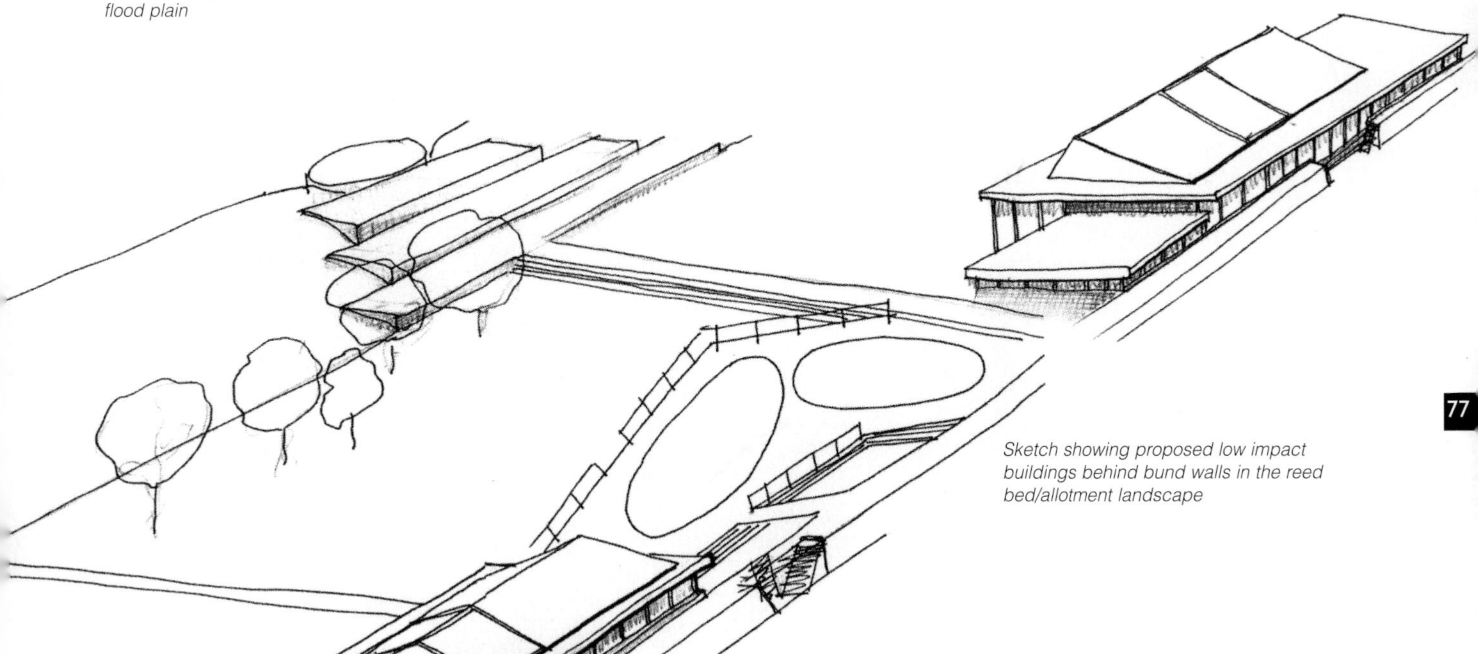

Sketch showing proposed low impact buildings behind bund walls in the reed bed/allotment landscape

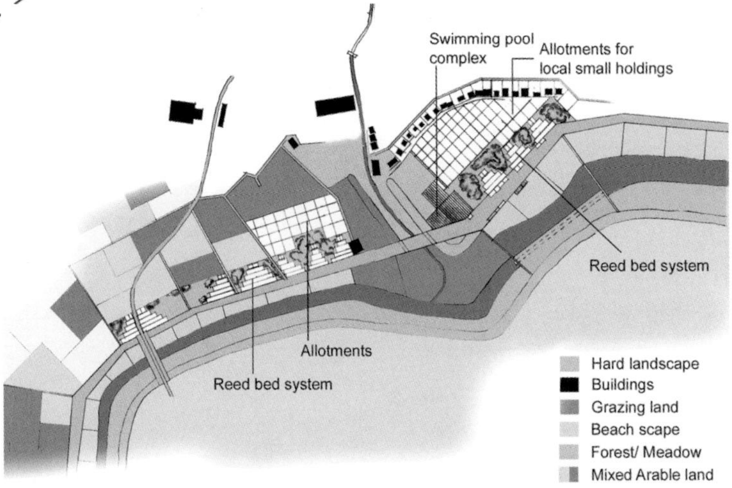

Swimming pool complex
Allotments for local small holdings
Reed bed system
Allotments
Reed bed system
Hard landscape
Buildings
Grazing land
Beach scape
Forest/ Meadow
Mixed Arable land

Location of scheme along river flood plain

Rozia Adenan

The long, straight, dusty backstreets, former haunts of terror gangs, are now thoroughfares for tractors and lorries. Tall courtyard walls barred with steel shutters close off the cosy world of home and look out on long straight barriers of barbed wire and mesh, topped with notices which scream KEEP OUT from barracks and marshalling yards. Between these walls, in hazardous chicken runs, children play.

Rozia assembled barricades of gabions, old tyres and scrap timber to carve out play zones for the children from the chicken runs. She encouraged the opening of the courtyards' steel shutters and the sharing of backyards, so that the space behind these bulwarks against the traffic also borrowed parts of the families' private realms. This space between the traffic and the house would become a place for children to prosper and explore, and for parents to gather, exchange gossip and care for their infants.

With a theme of therapy through protected play, she promoted the children's recovery from the psychological trauma of war.

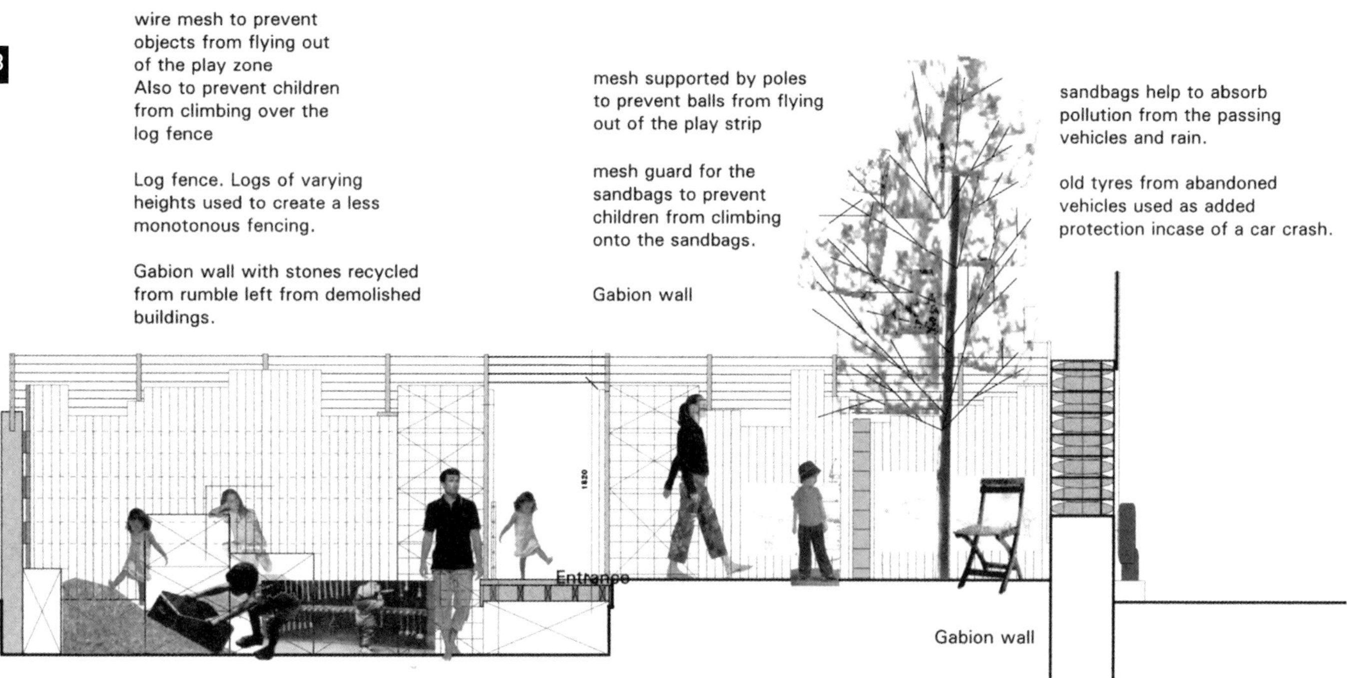

Rozia's protected zone for children to play

Below left: Boy's drawings often had a connection to war or to western hip-hop culture
Top right: Agon's drawing of two ghosts beside the dead body of his brother
Bottom right: Blenda's drawing of her best friend Pranvera

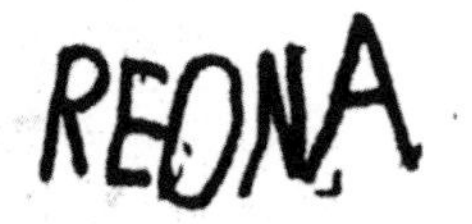

Top: Rita's home with one light bulb
Bottom: Reona, 6, as herself

4.6 The Bus Stop Bridge Area

On the north-west periphery of Vushtrri, there is a large, open, boggy, ownerless piece of ground where children misbehave, sheep roam and no-one settles. This provides the foreground for a calm reflective view from the Bus Stop Triangle over the 'Roman' Bridge (the oldest public artefact in the town) to the hills beyond. Set back from the rutted road that runs alongside the bridge is the timber clad Gymnasium, built by the Swedes and the site of a recent massacre.

The assembly of ancient stones is redundant. The river which it used to span had long ago been diverted to leave this flat damp terrain over which horses drag carts full of timber poles, in procession, to market. Nevertheless as a symbol of a more stable and meaningful past the bridge remains embedded deep within the psyche of Vushtrri's inhabitants.

Having traversed the Bridge Common, visitors enter Vushtrri proper via the Bus Stop Triangle, which is like a mouth clamped within the jaws of the town fabric. Here, returnees are disgorged from beaten up coaches onto the muddy surface of the triangle, kept squelchy by the diluted blood that bubbles up from cracked manholes as the effluent from the daily abattoir cull meets the inadequate drainage system. Climbing up to the cleaner pavement, home-comers are met by 'The Edge', a terrace of coffee shops and fast food joints, their first point of contact with the now resurgent town.

This area of investigation was the most popular with students, a rich and accessible source of inspiration. Ade and Namee made obsessive large-scale sketch maps showing the marks made by vehicles, pedestrians and animals throughout the day, like the ever changing marks on an ice rink. Ade named a jumble of under-occupied sheds and alleyways, backyards and allotments behind the Edge, the 'Hinterland'. Naomi surveyed the elevation of the Edge, identifying her site from an analysis of a schedule of dilapidations. Tom's library building reinforced the perimeter of the Bridge Common and celebrated the reassuring picture of habitual daily performance viewed from its windows. Georg wanted to transform the Common into an urban park, interwoven with a prairie of reed beds treating the district's waste water, to provide a swimming pool and public baths.

Adegboye Adetutu

Characterising the open space in this area as 'unkept' Ade chose to occupy or 'animate' pockets of this space using sound – to provide identity for individuals or groups. She asked, could a pattern of moments be defined that would enable these pockets of people to connect with each other so as to contribute self consciously to the growth of Vushtrri's contemporary public culture?

A soundscape of individual moments was drawn, mapping the sound fields that rose above the general background noise of the area. These varied in intensity and effect with the time of day. The moments included those generated by the flour mill, children's conversation on the bridge and the call to prayer floating above the noise of the transport interchange. By far the most overwhelming sound was traditional Albanian folk music, most noticeable on the streets, emanating from car stereos, tape recorders on makeshift stalls and music kiosks. From the Bridge Common through to the Triangle, moving into the centre of Vushtrri, Ade recorded the sounds of a demonstration organised to protest the continued detention of Kosovars in Serbia, including the sound of the names of the detained being read out over the tramp and shuffling of marching feet.

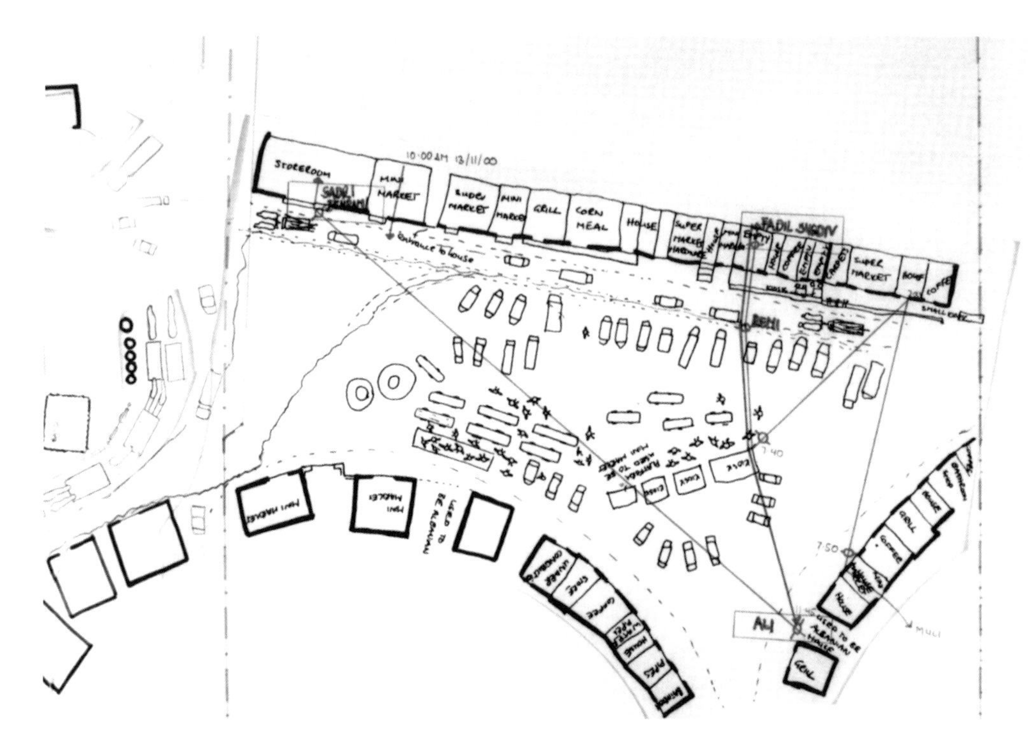

Sketch plan of existing bus stop area showing the Edge shops, buses, trucks, horse and tractor drawn carts and people

Drawings called the 'Score' were developed that classified sounds as being overheard, directional or intentional depending on the relationship between the source and its reception. Public expression and overheard conversation led to an idea about organised performance, recording and transmission.

On-site exercise

Ade made a breakthrough in her thinking by arranging to broadcast her own programme on local Radio Vushtrri. This followed a public presentation of the student's programme and was organised as a radio phone-in to debate the work. The programme was broadcast from a dilapidated building without water or mains electricity. Here the UN Interim Administration encouraged concerts, pantomime and theatre performances for all ages.

The proposal

Her proposal recognised the Edge as a significant threshold for the returning migrant and sought to include her Hinterland of redundant lots. She saw it as an existing social hub which could be adjusted to become a meeting area where new arrivals and local residents could meet during the daily chores and rhythms of their life. She suggested a sectional fragment of linked inhabitation that would serve as a prototype for this deepened Edge.

Stepping off the bus a mother and child might rest, eat and feel at home in an Edge café where the spoken word and live music might be overheard over tea. The level of privacy needed by each particular activity would be reflected in the sound transmission of the building elements. So, whilst on the first floor a women's centre offered confidential advice and employment, live music in the café would be recorded and broadcast from the radio station to the rear, set alongside theatre and concert facilities. The voices of women, children and minorities, whose voices would otherwise be lost, are included in the polyglot transmission.

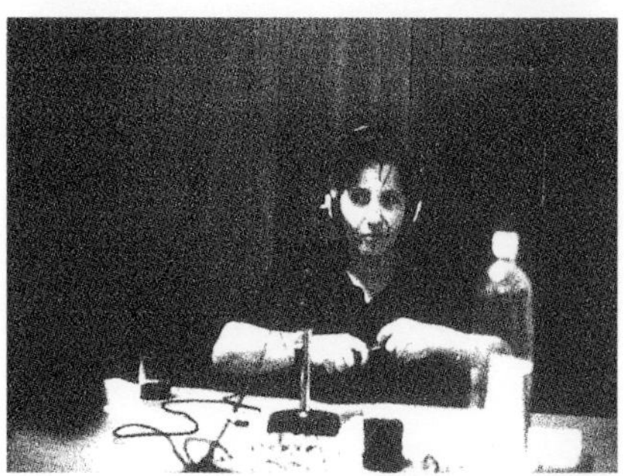

Top: Receiving an invitation at the bus stop to take part in a radio talk show
Middle: At the radio show
Bottom: Radio phone-in and chat about plans for Vushtrri

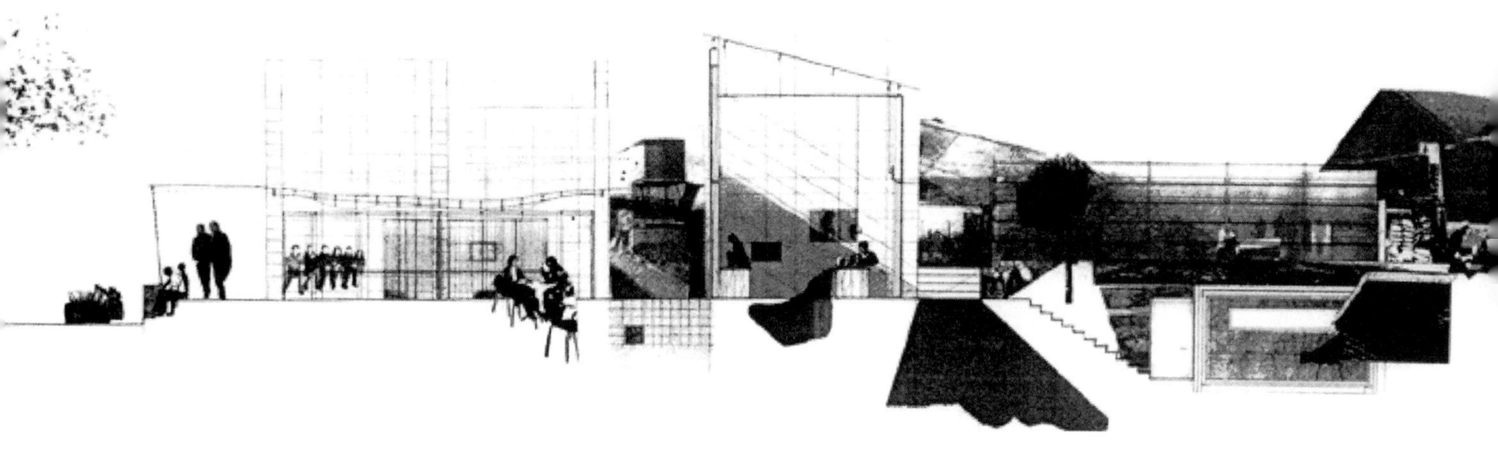

Collaged section through the Bus Stop Edge and Ade's Hinterland

Collaged section through pedestrian route between the Edge of coffee shops and bars and the Hinterland of public performance

Naomi Day

The Dive

Naomi's pilot project celebrated diving into Hampstead Ponds in winter. The experience involved changing (into a different identity), queuing and waiting (anticipation), the plunge (performance) and applause (as reward). By transferring this narrative to the Bus Stop area of Vushtrri, she aimed to articulate perceived public ambition to increase contact with the global community through sport. Naomi proposed to construct a diving pool where the Kosovo Diving Team would train for the 2008 Olympics.

Iconic image of diving above the Bus Stop Area

The view

Embedded in the façade of Vushtrri's gateway the public view from the diving tower and raised pool, over the water, Bus Stop and Common, to the hills beyond would be magnificent and relaxing. But far more upbeat, the iconic image of the somersaulting diver projected from the diving tower would be visible for miles.

The Edge façade

Naomi chose to replace a short-life shack (the least substantial building comprising the Edge) with her diving pool. The shack served kebabs and coffee, to people who passed by, on a pavement raised above the mud of the square.

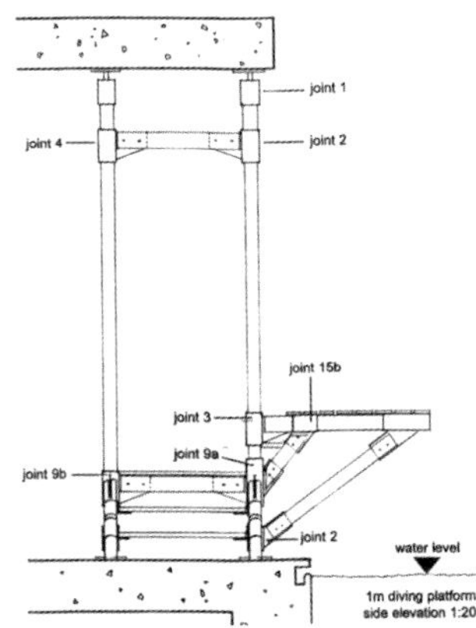

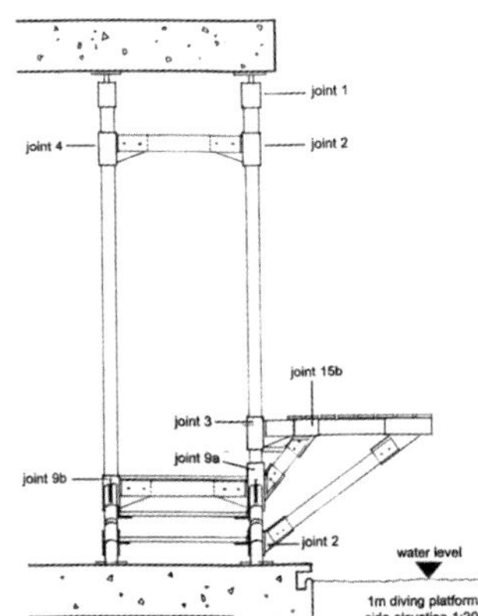

Top: Collage of proposed street elevation
Middle left: Detail of small diving platform using poles attached to concrete structure
Middle right; Section through the diving pool
Bottom: Existing street elevation to the Bus Stop Area

Naomi adopted and redesigned this public pavement as part of her scheme. The blank street side retaining wall of the raised pool would be adorned with posters and drinking fountains (as seen alongside Pristina's old mosque). Leading off the pavement would be a gaping pool entrance, controlled by a huge, locally made sliding steel door. Balconies from the upper floors protruding over the raised footway would provide views over the square to the hills beyond.

Technologies

The Tower celebrates height: built of reinforced concrete and floored using the FERT system it would collect water from surrounding roofs to top up the pool and support solar panels to heat the water. Local pole timber with metal connections would be used to construct stairs, diving platforms and a framework to support a quilted polythene covering for the pool in winter.

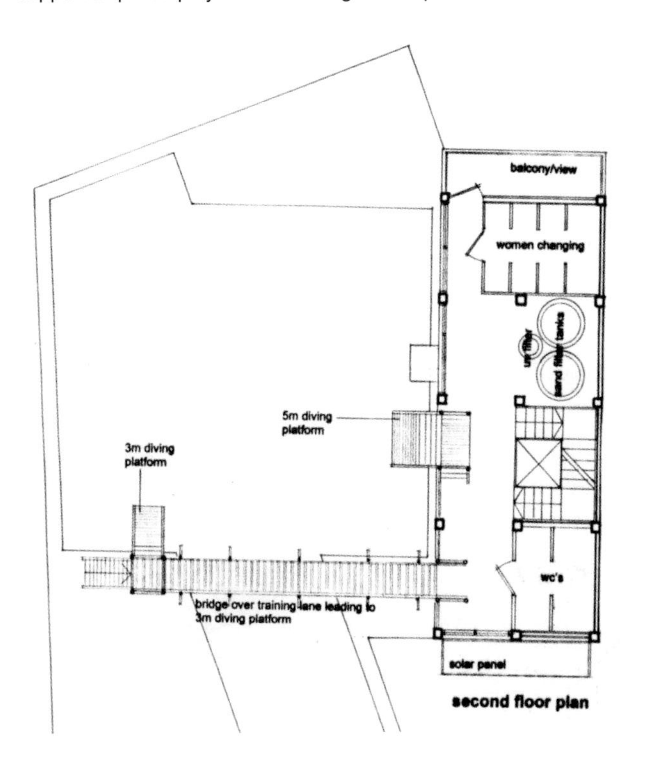

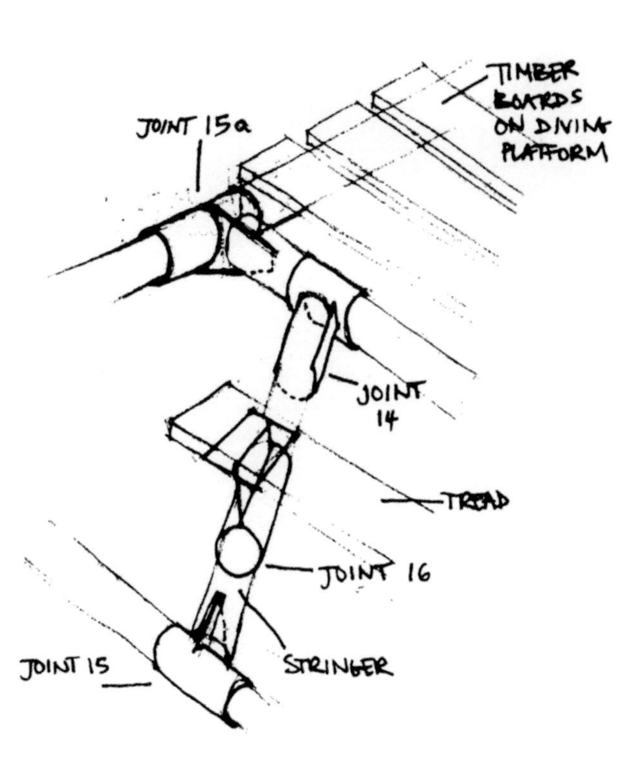

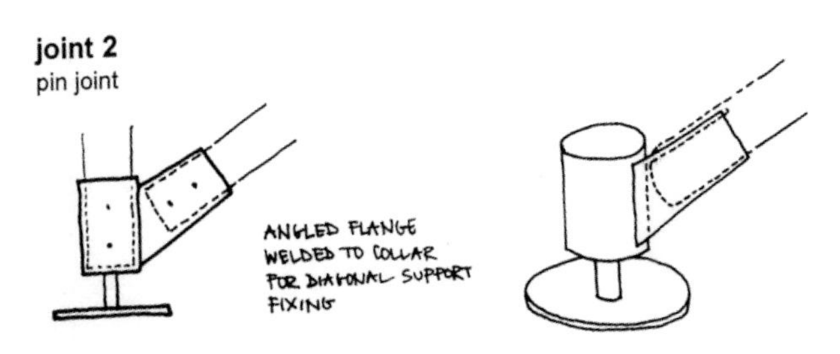

Above: Details of timber pole joints
Left: Second floor plan of the diving pool

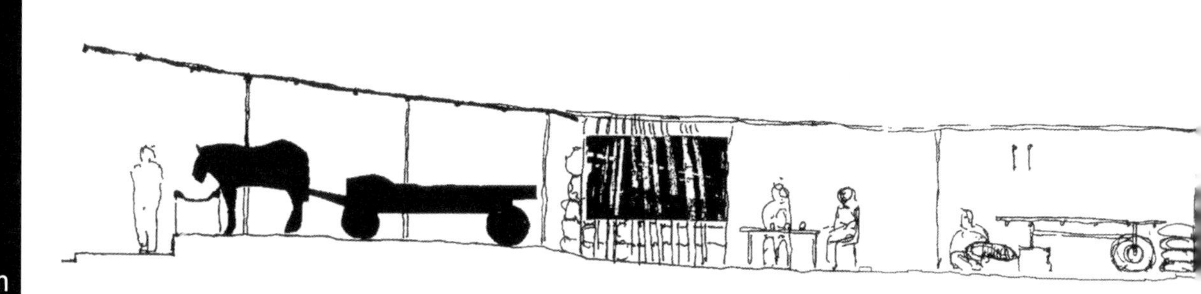

Namee Im

The journey

Namee made a journey back to the village on a horse drawn cart, which had delivered its load of firewood to market in Vushtrri. The tractor and trailer, loaded with family and possessions (symbolic of ethnic cleansing in TV reports of the war), was preceded by a horse and cart and followed by a taxi.

The survey

Namee surveyed a makeshift facility for stabling and feeding horses, resting and feeding their drivers and repairing harness and trailer, whilst preparing to return to the village. This facility was located at the nexus of Common and Triangle, a potentially valuable urban site. However, structures were ramshackle and constantly changing. It was perceived that it was unlikely that these workshops and hostelries would survive development and stay in the same ownership unless capital was invested in a careful plan.

Top: Existing stable, store and workshop
Middle: Proposed pole stable derived from modelling at CAT
Bottom: Proposed stable structure derived from Kosovan vernacular

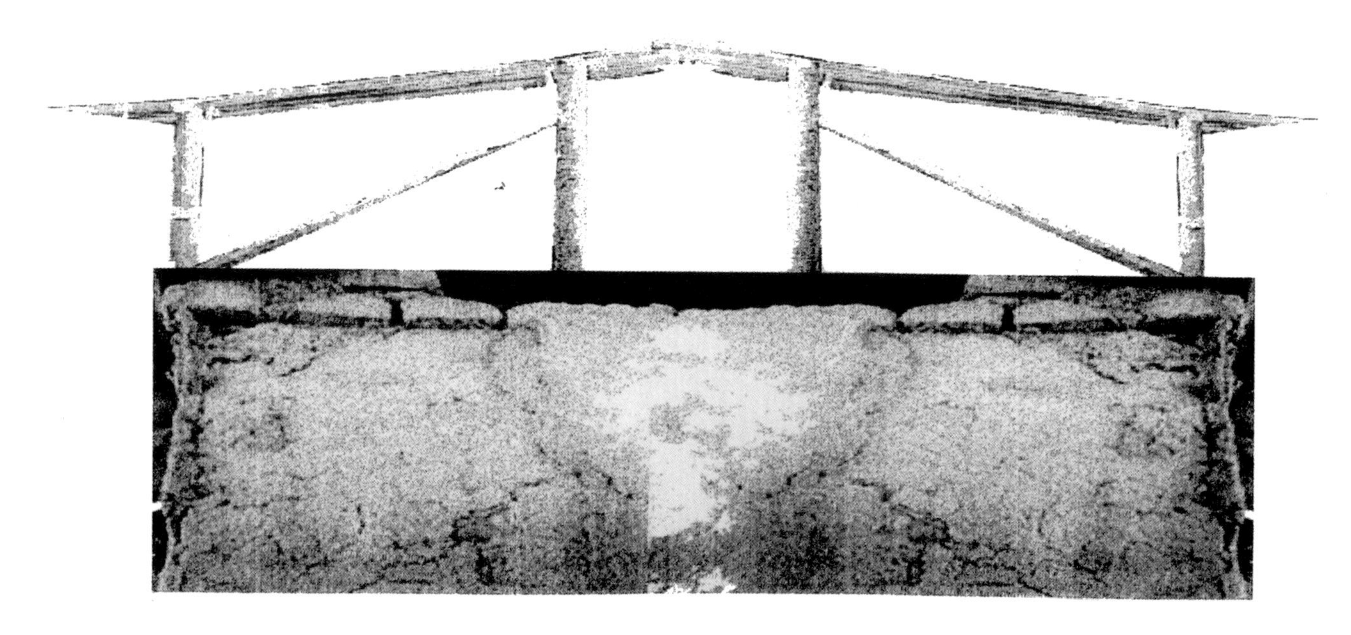

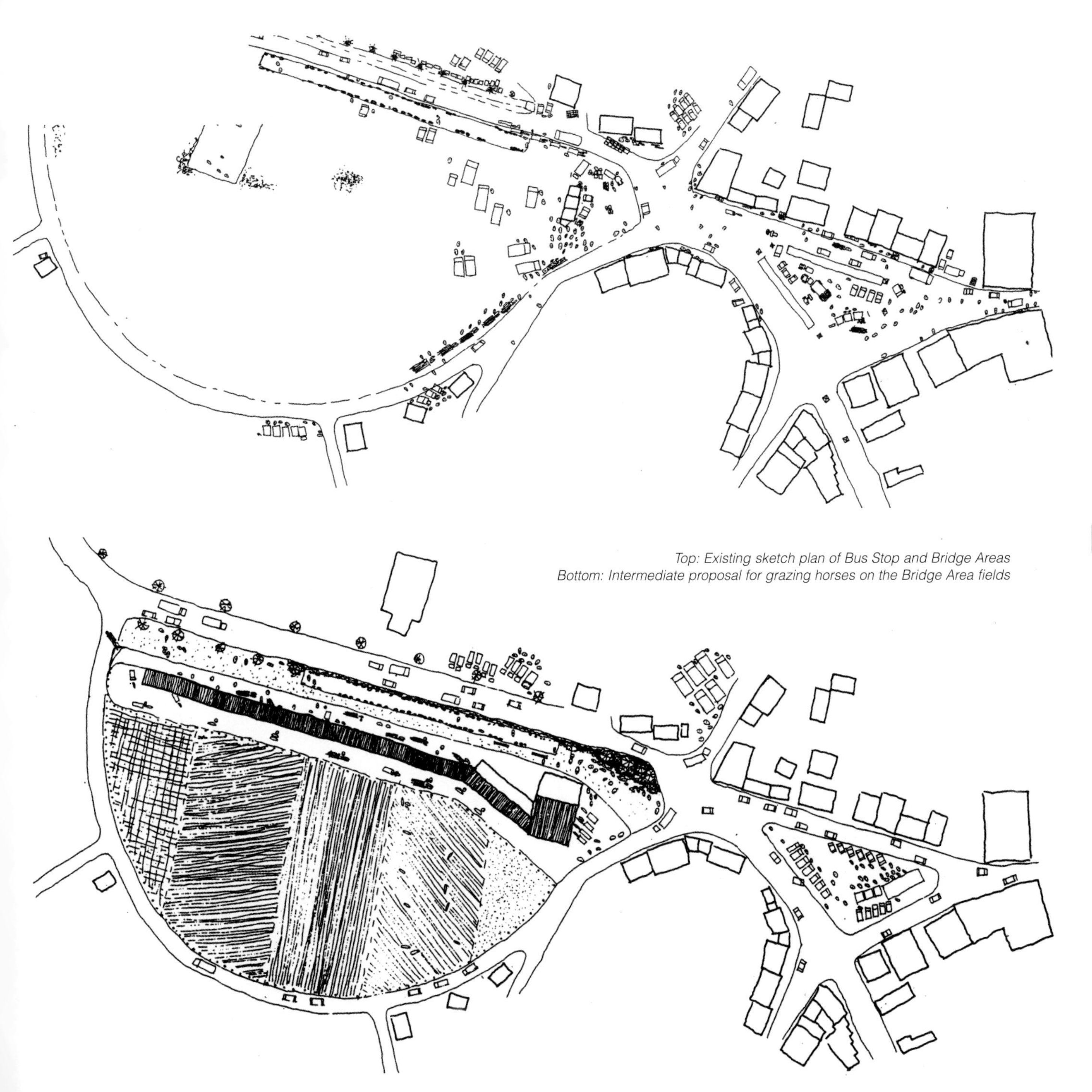

Top: Existing sketch plan of Bus Stop and Bridge Areas
Bottom: Intermediate proposal for grazing horses on the Bridge Area fields

The proposal

Namee choreographed the likely changes from stables to workshops, from rest and recuperation to hostel and hotel accommodation. She posited that if the changes could be foreseen and planned for, then ownership of workshop, stables and hostel by drivers for drivers could be maintained in the long term. She envisaged a form of empowerment, perhaps in the form of a driver's co-operative, through planning ahead. At first, using the empty common land to collect water and grow fodder, the emphasis was on cash-less self-sufficiency. Later, however, land was expected to increase in value, the Bridge Common would no longer be exploitable. Each proposed change was small but the capital invested for each phase would contribute to a larger pre-planned scheme – for a comprehensive workshop and hostel facility to be constructed one floor at a time, whilst still in use, as and when capital was acquired.

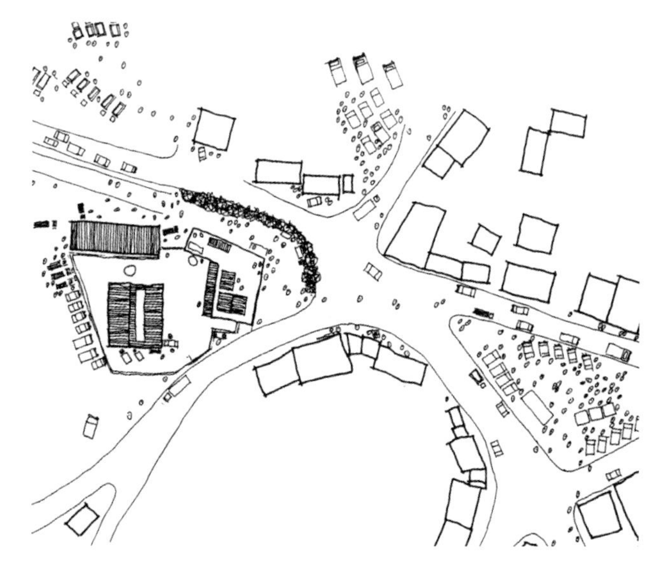

Proposed sketch plan of drivers' hostel

Top: Existing elevation of shops and workshops on site of new hostel
Bottom: Proposed elevation of drivers' hostel and mechanical workshops

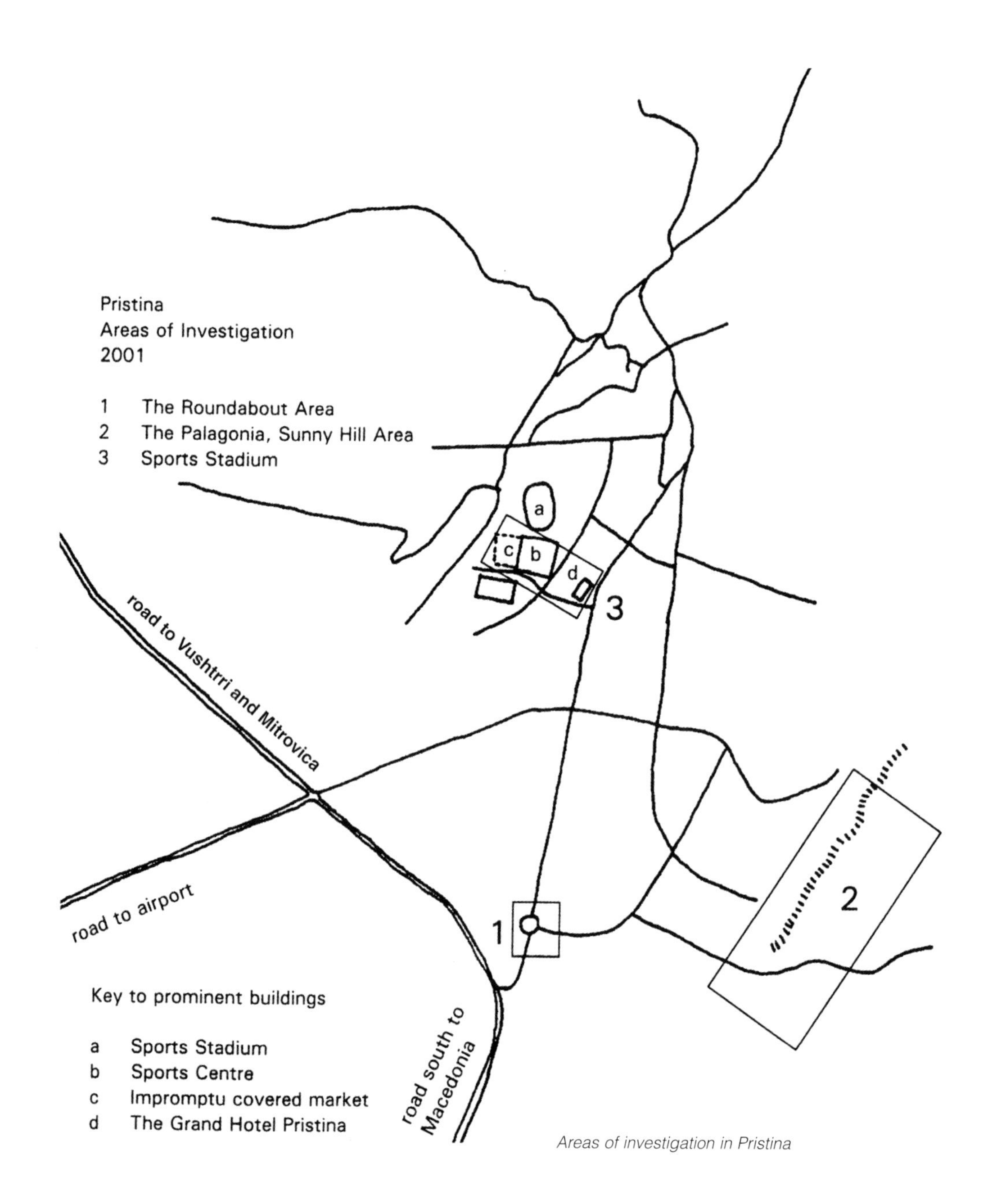

Areas of investigation in Pristina

Pristina

Top: The Roundabout from a vacant slot in the Archigram block
Middle: The warehouse from the Archigram block
Bottom: The occupation of a slot in the Archigram block

4.7 The Roundabout site

All major roads to the south of Pristina meet at the roundabout. If you drive from Macedonia in the south, Mitrovica to the north or straight from the airport you will pause at the roundabout. Marking the start of the city, the forbidding ranks of tall grey tower blocks, holding a pall of smog at their tops, will give you pause for thought before you plunge into its depths. If you are a paying passenger in a shared taxi, van or minibus you are likely to be dumped at the roundabout. Here you will be offered a wide choice of services (food, music, hair cut, clothes) sold from illegal kiosks, varied in shape, size and colour, which perch opportunistically at the side of the road.

On the north-west side of the roundabout is an imposing eight-storey reinforced concrete skeletal frame. Students called it the 'Archigram' block (after the notion of 'plug in city' invented by that group). Construction was commenced by the Serbs with the intention of housing Serbian refugees from Bosnia. After the Serbs left, with only the raw frame of columns and floor slabs complete, internally displaced Kosovan Albanians began to squat and complete the structure. Less than three years after the war, the building is now three quarters occupied and temporary licences have been granted for mains connection to water and electricity. New occupants add walls, doors and windows and the façade is now a piecemeal amalgam of brickwork, blockwork, timber and tiling.

To the west of the roundabout is a complex of empty, mostly single-storey, sheds, occasionally used for car repair, named the 'Warehouse' by Angels and Jean. The largest shed is steel framed with a north-lit, saw-toothed roof; the others adjoining it are smaller steel portal or concrete frame structures. Contemporary Kosovan building structures are totally reliant on reinforced concrete. Any proposal for new steel frame construction would need imported steel. Redundant steelwork with its capacity for deconstruction and re-erection in another form is an opportunity for recycling which the students latched on to.

To the south of the Warehouse is an open, flat piece of land with a redundant petrol station at its southern tip.

On first inspection, there was bad feeling between the legal occupants of surrounding government social blocks towards the occupants of the Archigram block and the kiosk owners. They were seen as incomers, taking advantage of the uncertain legal position regarding connection to services and ownership of land and buildings. The long term occupants expected the new authorities to invest in major new transport infrastructure and expected the under-occupied land to either be subsumed in a road building scheme or absorbed within a 'green belt' of parkland, which they would like to see circle the city.

Angels and Jean chose to take the part of the incomers, to investigate the implications of an expansion of their ambition to exploit land that was blighted by lack of investment. They would act as facilitators to the incomer community, working to ensure that expansion of local entrepreneurial ambitions could piggyback on any new road scheme whilst still providing social and public space in the interstices.

Jean Dumas

Jean proposed to act as a facilitator to the kiosk owners and internally displaced refugees occupying the Archigram block. His proposal illustrates that, even with minimal capital resources, there are a wide range of economic, material, and spatial opportunities available over a 20 year period if the community is willing to co-operate. His scheme focuses on adapting the Warehouse structures to provide workshops, cinema, theatre and studios for people to make, learn and teach. He shows how the large expanse of shed roof could become both a generator of energy and a rainwater collection device.

Workshops

Residents of the Archigram block included an unemployed musical instrument maker and other craftsmen whose creative abilities were being wasted through lack of opportunity. They would now be able to practise their individual trades in small workshops within the rehabilitated shed. On a slightly larger scale, with the technology used to manufacture kiosks (plastic, aluminium and steel) the workshop would make insulated cladding panels to help complete the façade of the Archigram block and other similar empty structures.

Energy

We visited Pristina three years after the war and despite staying at the 5 star 'Grand Hotel' we still experienced strict rationing: no water from 12am to 6am, no hot water in the afternoon and intermittent power cuts of an average two hours duration, two or three times a day.

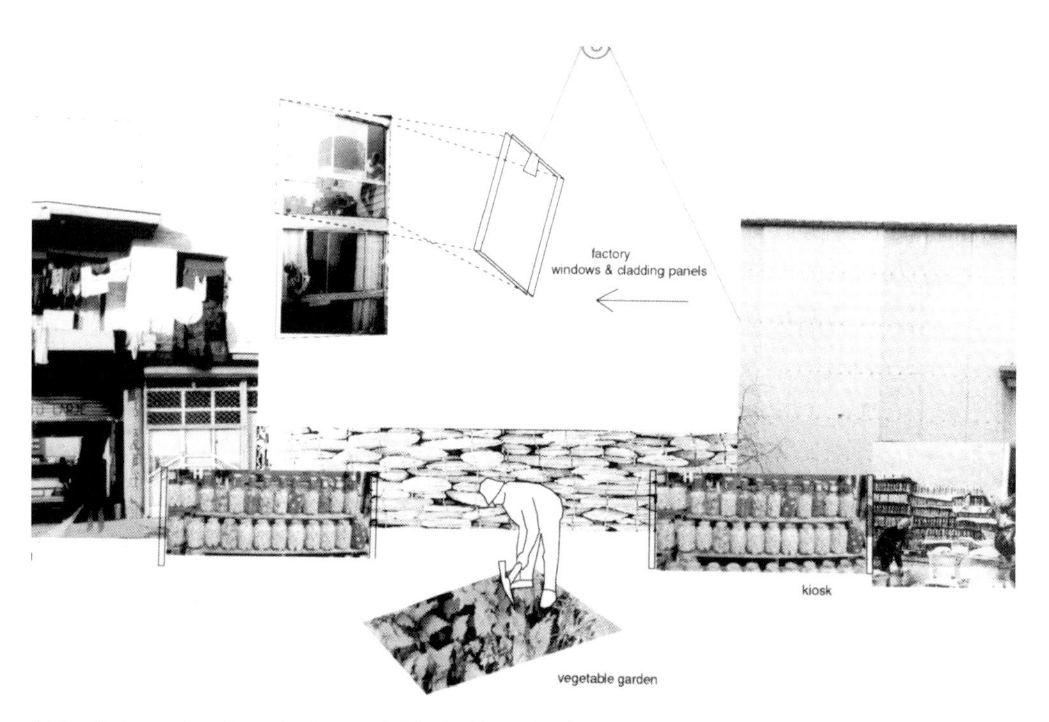

Using kiosk manufacturing technology to clad the Archigram block

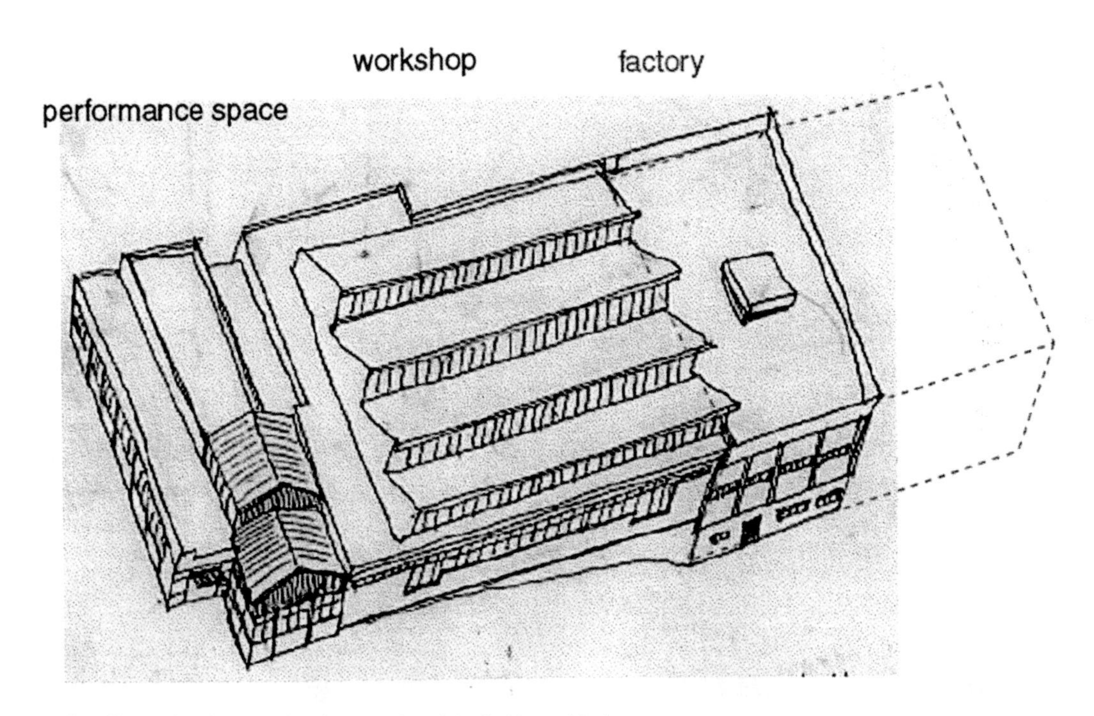

Top: Sketch showing completed occupation of the Archigram block
Bottom: Sketch of existing warehouse showing strategy for rehabilitation as performance space, workshop and studios

Top: Corner of warehouse as existing, allocated for theatre space
Middle: Phase 1: strip off roof, save portal frames, windows and wall cladding
Bottom: Phase 2: space left for theatre and foyer identified

The southern slope at the back of the north facing shed roof was ideal for installing solar (heating hot water) and photovoltaic (electricity generating) panels. But could the use of imported materials and components be justified to facilitate the sustainable, pollution-free collection of energy when changes to existing buildings were to be made only of local and recycled materials? Jean thought that energy was in such short supply and its current production so polluting that investment in solar panels was justified. Making simple solar panels from blackened recycled radiators set behind glass and using a local method of encapsulating imported silicon cells would reduce the capital cost.

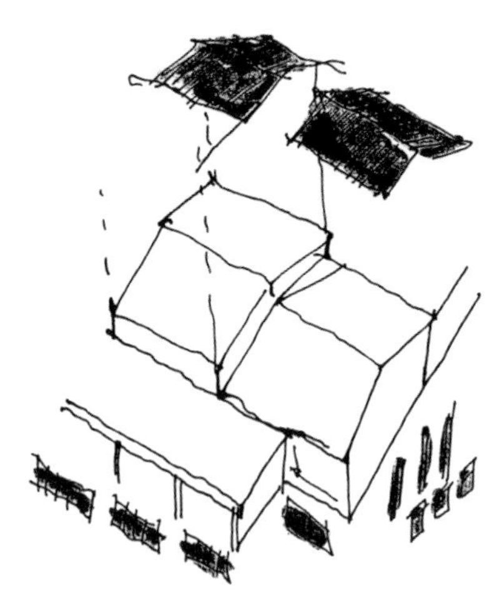

Water
Car ownership increased tenfold in the years following the war, crowding the roads and increasing sales at roadside kiosks. At the base of the Archigram block a new roadside car wash had just been installed: the latest craze in roadside economic activity. All that was required to set up in business was a water supply close to the road and investment in a generator and pump, attached pipe and sponged nozzle. However, the mains supply was intermittent and car washes were now competing with domestic needs. Jean's proposal included the collection of rainwater from the large expanse of shed roof. Ferrocement filtration and storage tanks were strategically placed at collection and distribution points so that the Roundabout community could top up the inadequate mains supply to their car washes, kiosks, workshops and places of entertainment.

Social space
The north-east corner of the Warehouse shed complex would be developed into a foyer and theatre. Between the foyer and the Archigram block Jean planned a public square. The existing small steel shed would be dismantled and reassembled on top of a raked concrete seating base. The double height foyer would serve the theatre and give access to a roof restaurant. Cladding would be of glass, and aluminium panels made in the workshop. Entrance to the foyer would be through large sliding doors made from recycled metal sheets.

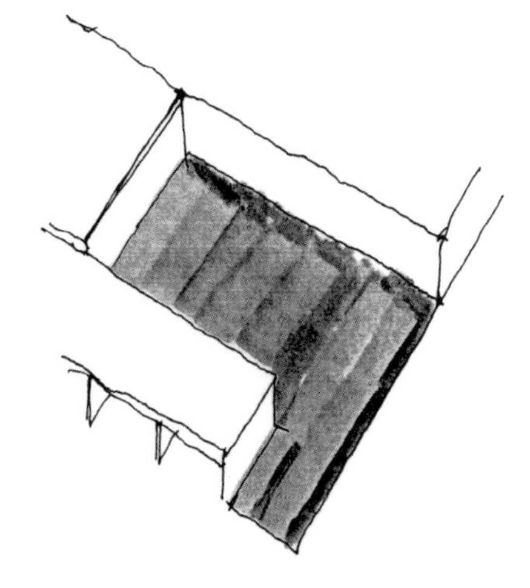

Top left: Phase 3: add mezzanine floor at front with new reinforced concrete frame giving access to roof terrace
Top right: Refit steel portal frames to provide sloping theatre roof
Middle right: Add front foyer with lightweight timber scaffolding frame clad with recycled steel sheets incorporating reused large sliding industrial doors opening onto public square
Middle left: Elevation of new theatre within the warehouse
Bottom: Long elevation of workshops and new theatre, section through public square and Archigram squatted block

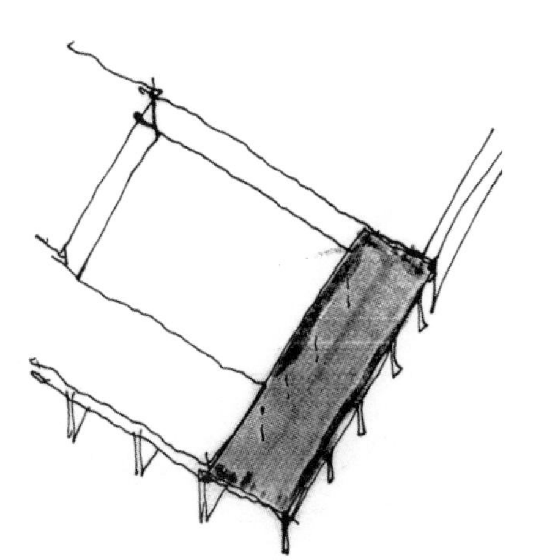

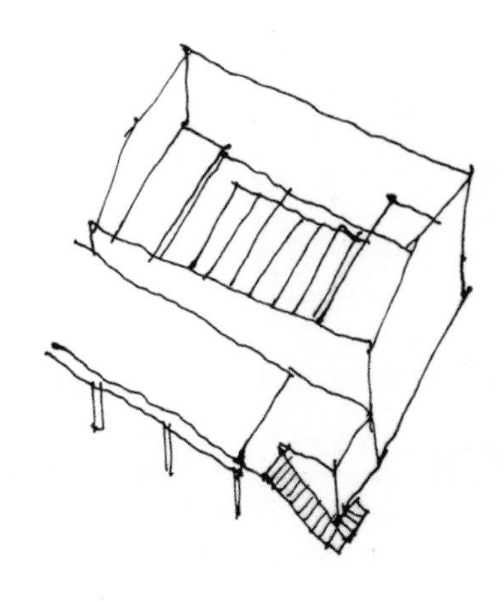

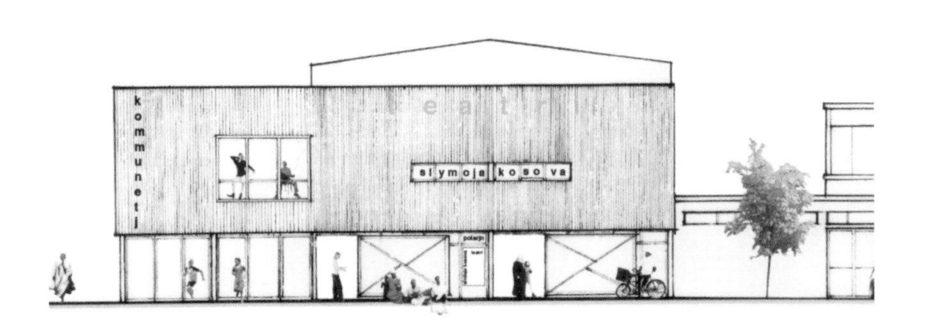

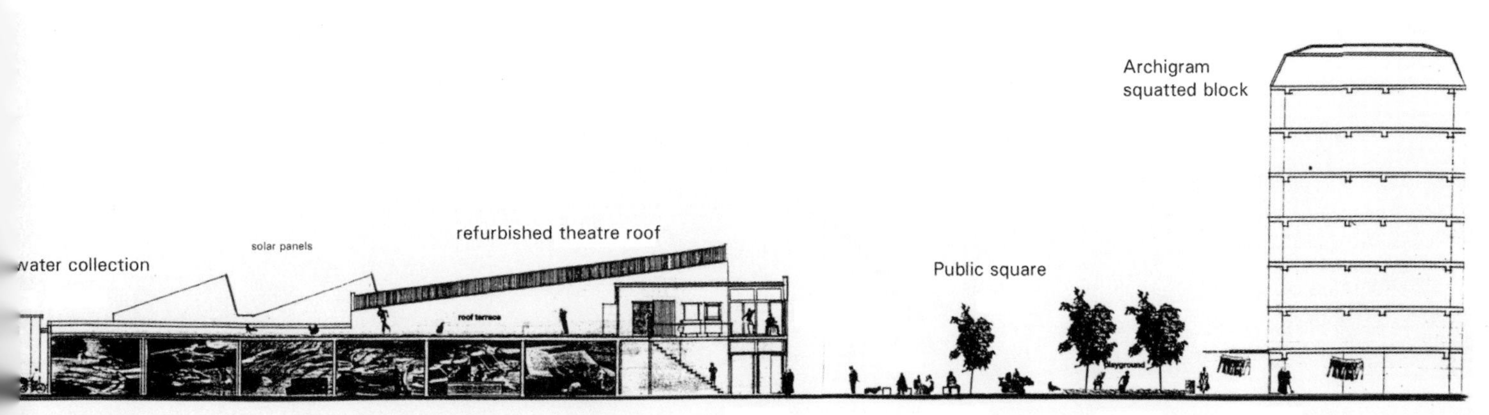

Angels Lopez

Shelter in a shed

Angels took her cue from the floods of internally displaced people moving to the city of Pristina to find work. For these people the Roundabout site was a good place to look for, at least, temporary accommodation. The Archigram block showed how, with self-help methods of construction, redundant structures could be occupied in a permanent way. At the other end of the scale, a new hotel had been built on the north-east corner of the Roundabout, reputedly more luxurious than the Grand Hotel. Soon there would be an increasing pressure for land at this, Pristina's gateway. Angels' approach was to focus on turning part of the Warehouse block into shelter for the poorer, recently arrived migrants using the available sustainable loose-fit technologies. Over twenty years she hoped to promote a continuously improving infrastructure for dwellings in the area occupied by the sheds closest to the Roundabout.

Her first modest intervention was to provide a roadside hostel/motel located in a single-storey concrete shed with Portakabin overspill accommodation placed on part of the flat shed roof using kiosk technology.

Two perspectives of accommodation for internally displaced people constructed within the recycled framework of the warehouse

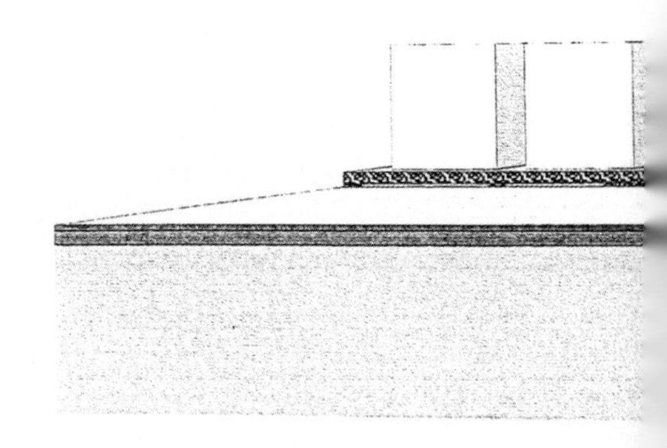

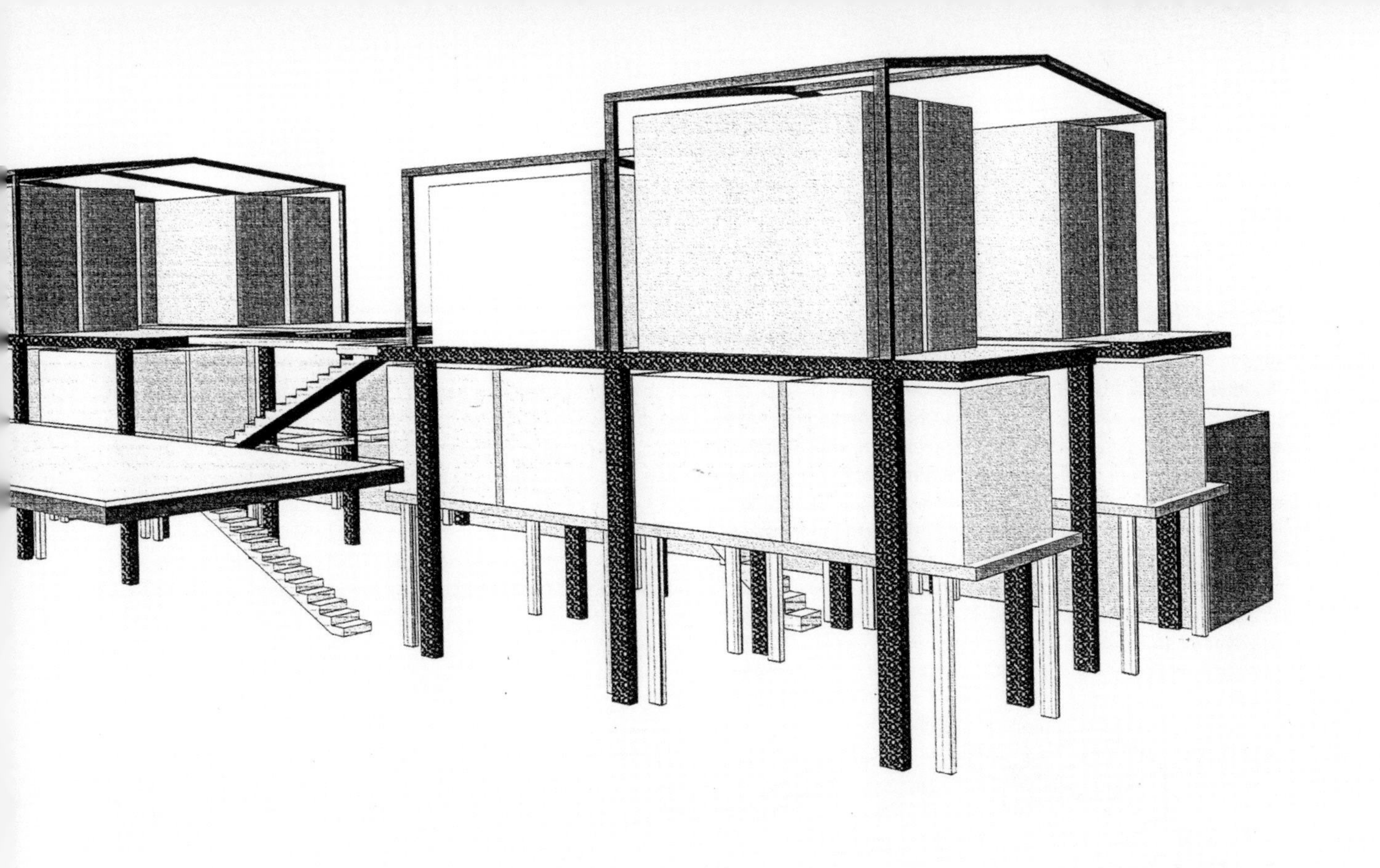

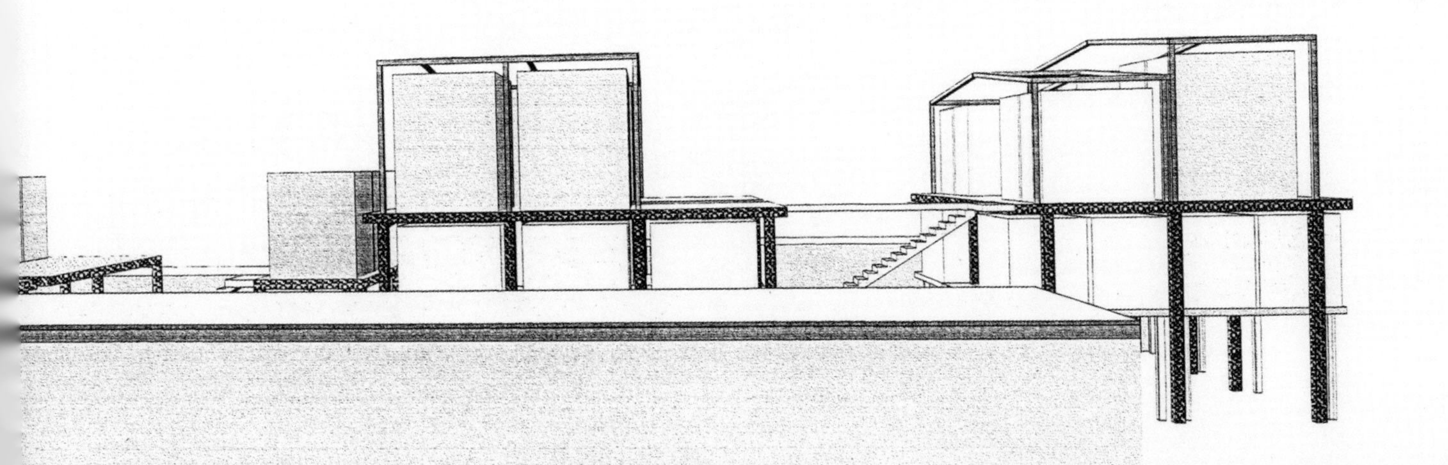

Angels was particularly struck by the cost effective shed accommodation offered to asylum seekers by the Red Cross at Sangatte in France, many of them Kosovans. In Sangatte both tents and Portakabins were placed inside the shed with the intention of providing privacy. However, conditions were overcrowded, insanitary and cold. Natural light was minimal. Anxiety and boredom were high, as was turnover. There were no gaps between the camp beds crammed into the cabins and Muslims were forced to pray in a corner of the wash room. Angels was sure this could be done better at the Roundabout site: by limiting numbers, by giving the occupants an opportunity to 'own' their part of the building and by encouraging residents to initiate improvements to the fabric of the building.

Angels proposed a phased three-storey framed construction infilled with family cabins. She would cannibalise the existing sheds. The recycled steel portal frames were placed on top of a double-height reinforced concrete table structure. Slipped under the table was a single-storey timber pole framework built by self-help. Natural light and ventilation entered the complex through open courtyards and central wells. As in Jean's scheme the adjoining saw-toothed roof became a source of solar heating photo-electric lighting and rainwater collection. There was a looseness to this basic strategy that would allow the final construction to evolve in response to the efforts of the occupants.

02

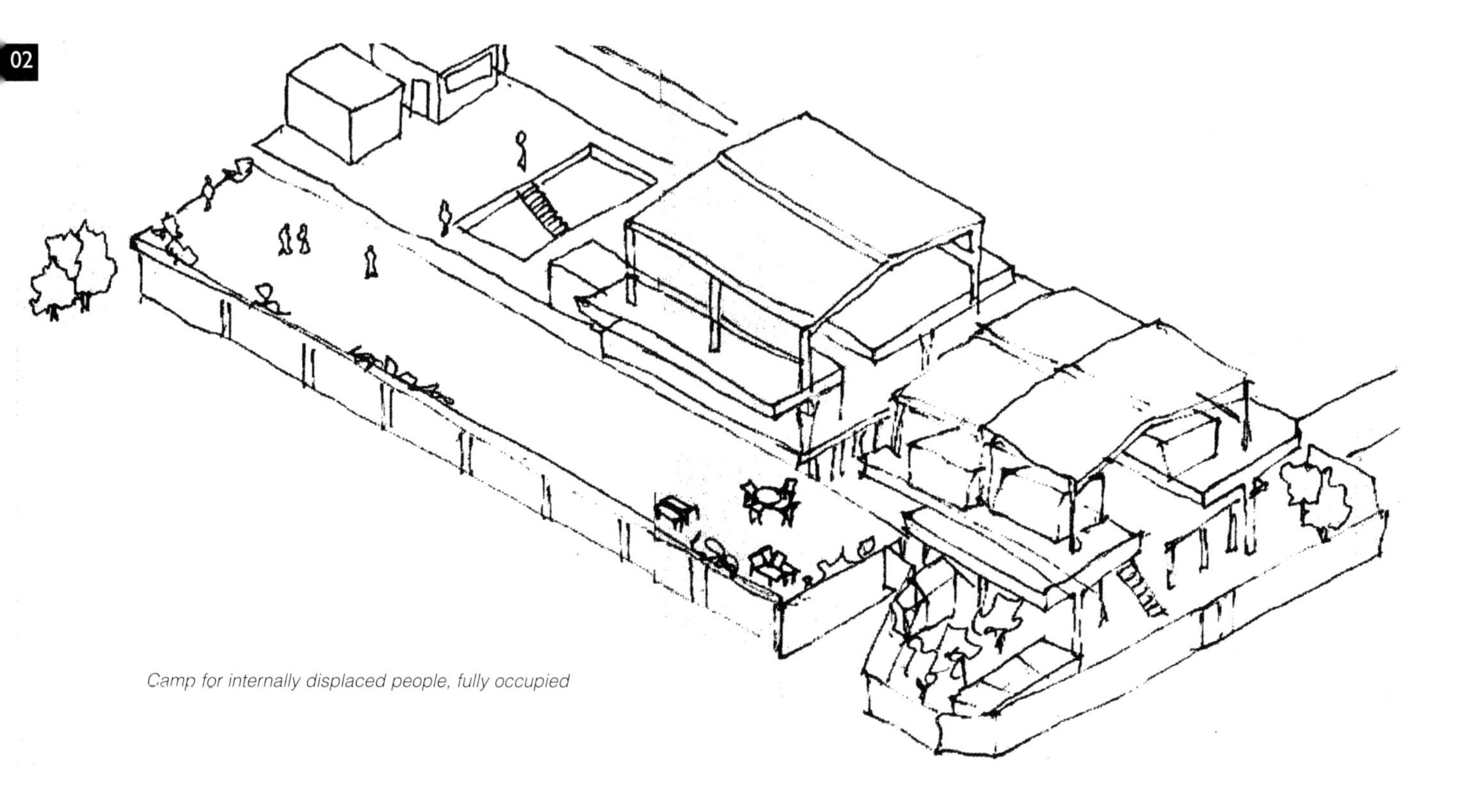

Camp for internally displaced people, fully occupied

4.8 The Palagonia, Sunny Hill area

Slope and sun

The squatters of Palagonia, on the outskirts of Pristina, are unlike those in the favellas and bidonvilles of the Third World. Neither wretched nor impoverished they have illegally occupied individual plots of land and built detached family houses with their own individual Calor gas-fired boilers for domestic heating and hot water.

The soft meadowland rolls down from these settler plots to a natural watercourse and then rises steeply to Sunny Hill, a neighbourhood of State-owned social housing notable for its plethora of satellite dishes clamped to the outside of the multi-occupied, multi-storey blocks. Here residents are poor but legal, without private open space and dependent on an intermittently functioning, Soviet-era district heating system. This inefficient plant runs on electricity generated from poor quality coal in the notoriously polluting Oblic power station. In winter, power and heating cuts occur daily as government engineers struggle to share the inadequate energy supply equally amongst the different urban districts.

The watercourse itself is invaded by effluent from cracked, under-capacity concrete sewerage pipes. Daily trash from Sunny Hill is tipped down the steep slope. Only an impromptu, rickety plank bridges the stream allowing children from Sunny Hill to join their cousins for a game of football on the flatter Palagonian meadowland.

Both communities use the nearest shops and hospital, both of which are in Sunny Hill and both lack recreation space for children and young people. Both communities would benefit from schemes using the valley as a park to zip them together. The south-east facing Sunny Hill slope offers panoramic views and the opportunity for tapping the sun's energy. The lie of the land provides an opportunity for rainwater collection. At the moment, water pollution literally turns people's heads away from seizing these opportunities (due to the smell).

View from Sunny Hill over the site to Palagonia

Lole Mate and Tim Wong

Both students imagined the steep Sunny Hill slope arrayed with glinting photovoltaic and solar water heating panels, serried ranks of water tanks and displays of colourful clothes hung out to dry in designated clothes washing areas, in among a few terraced pavilions. Paths would meander up and down between flowering reed beds and the occasional ramp would lift the visitor up from the new pedestrian bridges below, from terrace to terrace like steps in a game of Snakes and Ladders.

Plan showing integration of park with water treatment, hammam and leisure pool

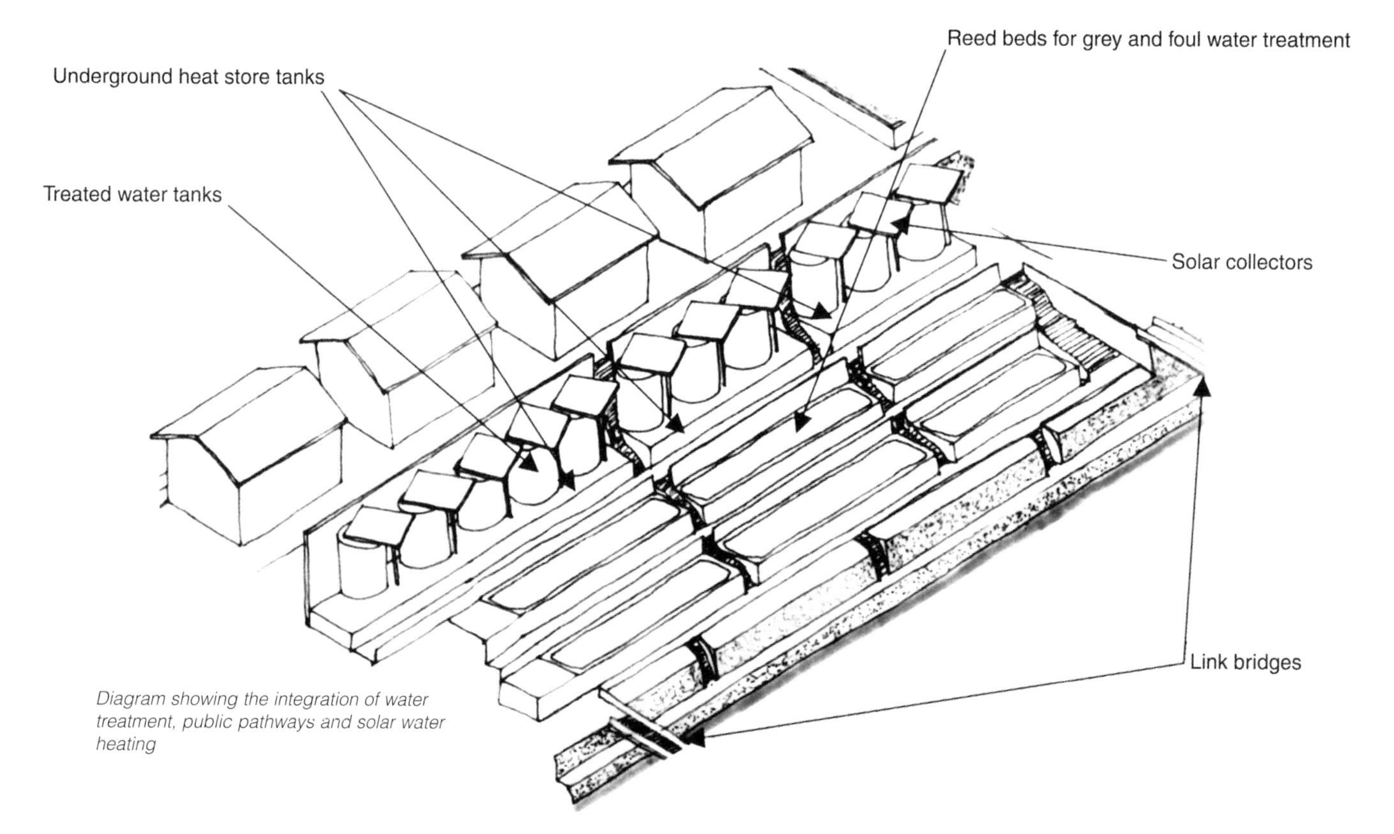

Diagram showing the integration of water treatment, public pathways and solar water heating

Tim proposed that one of these sparse, long, narrow terraced pavilions would be accommodate a hammam, swimming pool and laundry and would also allow for relaxation with a view. Whether this involved splashing a little, folding washing, leaning on a rail and pausing, or stretching out on a grass roof, the long view could, because of the stepped section, be enjoyed in private – either as an individual or in a gender or age defined group. He managed the steep hillside by alternating horizontal bands of reinforced concrete with steeply stepped, gabion retaining walls. Sturdy timber pole portals supported publicly accessible turf roofs over the pavilion pool and hammam. Lower down, nearer the stream, promenade decking was perched off a variety of slope stabilising structures.

Lole concentrated on turning the steep slope into a public park, a sewage treatment plant and a district power plant all rolled into one. Public laundries supplied with warm water directly from solar heated water tanks would be available for family washing with a view. The reed beds and associated planting would attract wildlife, driven away by pollution, back into the area. A planted river walk would provide a longer recreational stroll connecting to other parts of Pristina's regenerated peripheral green lands.

Section through swimming pool and open air laundry spaces

Site investigation and modelling

Research on site involved extensive measurement of the landscape to understand its patterns and potential. At the Centre for Alternative Technology (CAT), in Wales, a large-scale model of the watercourse and its valley was constructed from measurements taken on site. This enabled the students to understand the scale of the valley and judge the appropriate sizes for elements which might be inserted within the landscape.

Precedent

Working prototype examples of rainwater collection and filtration, water tank construction, reed bed treatment of grey and black water, photovoltaic and solar water heating panel technology were all explored during a workshop at CAT. The specific example of district heating used at Lyckebo in Sweden became the basis of both students' proposals to light and heat at least 550 family dwellings using photovoltaic electricity generation and solar water heating.

Concept photograph showing the conviviality surrounding the family washing activity on Sunny Hill

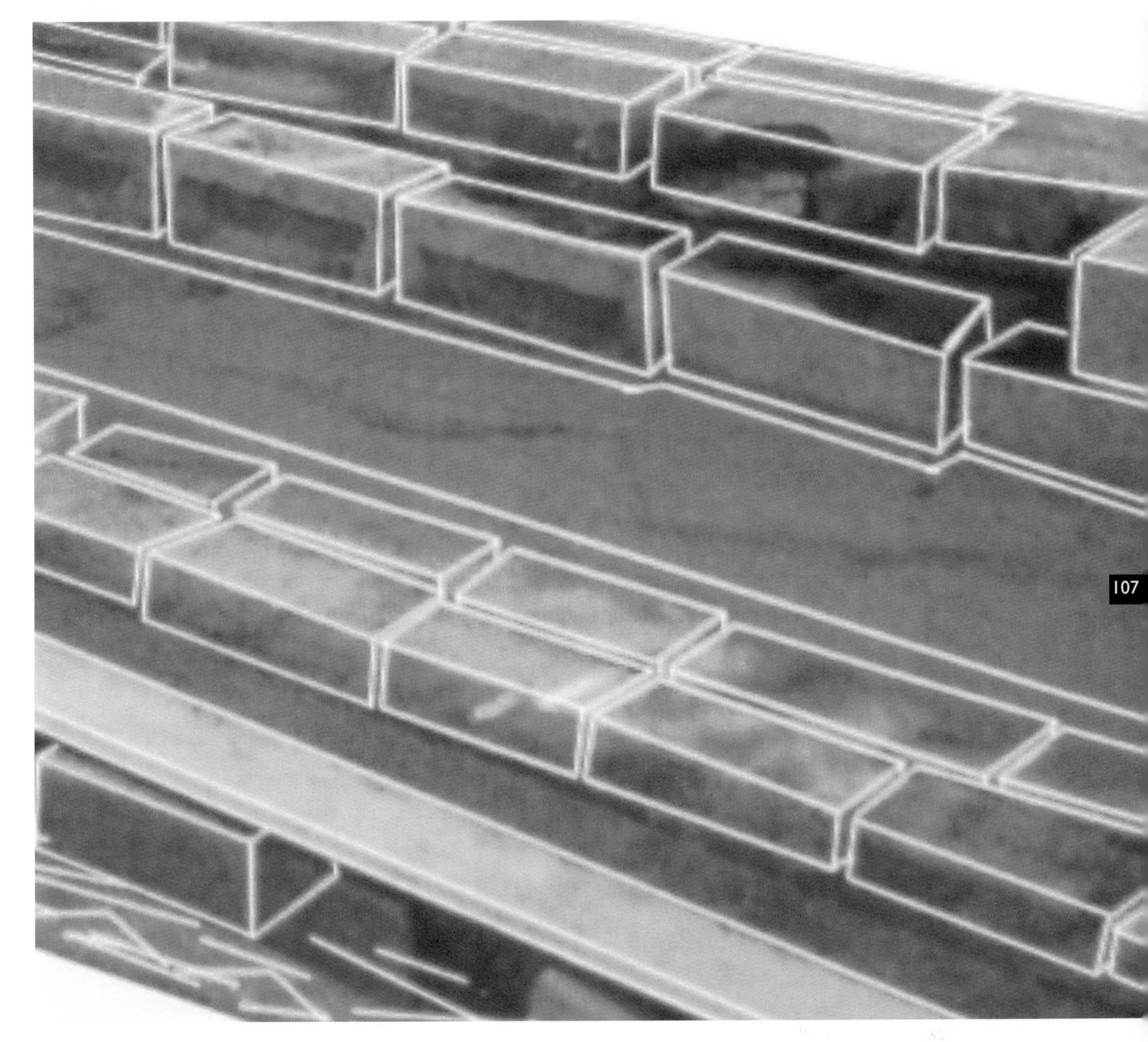

The build up for a large scale contour model of the site used as the basis for Lole's design idea

Model of proposed terraced park and reed beds

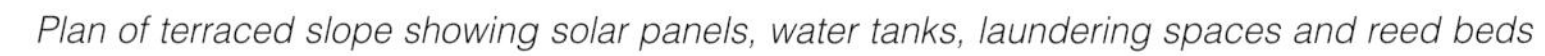
Plan of terraced slope showing solar panels, water tanks, laundering spaces and reed beds

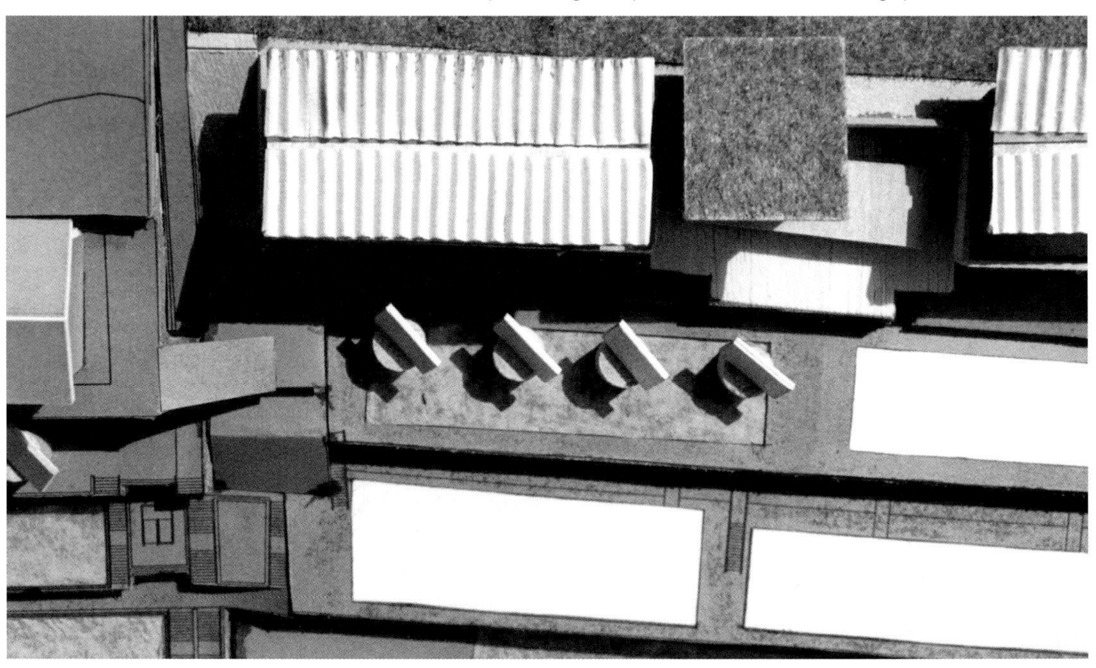

4.9 Sports stadium area

In the sixties, to maintain the 5 star status of the Grand Hotel in the centre of Pristina, it was decided to connect it to the 'Palace of Sport' some 500 metres away. Following an international competition, a huge reinforced concrete, deck access structure surmounted by a long span steel roof was designed. The concrete megalith was constructed in the seventies but without deck access to the hotel. The hotel and sports centre survived the war (despite NATO's precision bombing of the nearby police 'interrogation' centre), only to be devastated by a gas explosion a year later.

You can look at, and out from, this striking example of central city redundancy. Viewed from below, standing in a neighbouring field of coloured plastic bags waving on stalks, the contrast between the brightly painted informal kiosks lining the edge of the site and the empty concrete backdrop is stark. From above, perched on the uninhabited deck, a wide panorama of blue and red UN plastic roofs covers an impressive sprawling impromptu market, teeming with consumers.

Pristina is now an international city with a large foreign presence (with some 60,000 UN and KFOR personnel alone). However, Pristina is also characterised by ethnic enclaves separated by fences and suspicion. The covered market appears to present the greatest opportunity for social mixing. If only some of this opportunity for intermingling could be extended to civic and leisure activities... This notion generated much debate with our Pristina hosts.

Magda Raczkowska

Magda was inspired by the sea of polythene covering the market to produce a design idea focused around a roof. She cleared out the rubble from the Palace of Sports, identified a pedestrian route from the hotel to the impromptu market, via the burnt out concrete shell, and recycled the skeletal steel roof to provide a community canopy. She encouraged the market to spread under the new canopy mixing this with cinema, theatre and other civic and leisure activities.

Whilst the community canopy covers the upper concrete deck, it follows the geometry of the market roof, rather than that of the deck, providing a range of dynamic spaces as the two grids overlap and collide. The twisted geometry of the roof grid over the raised deck is mediated by a further horizontal layer – the new floor overhanging the deck in places to provide access to the street below and transform the elevation. Beneath the new floor are services and a transfer structure. Under the canopy, walls and partitions are to be built from straw bales, or other recycled or scrap materials, echoing the social and material dynamism of the impromptu market.

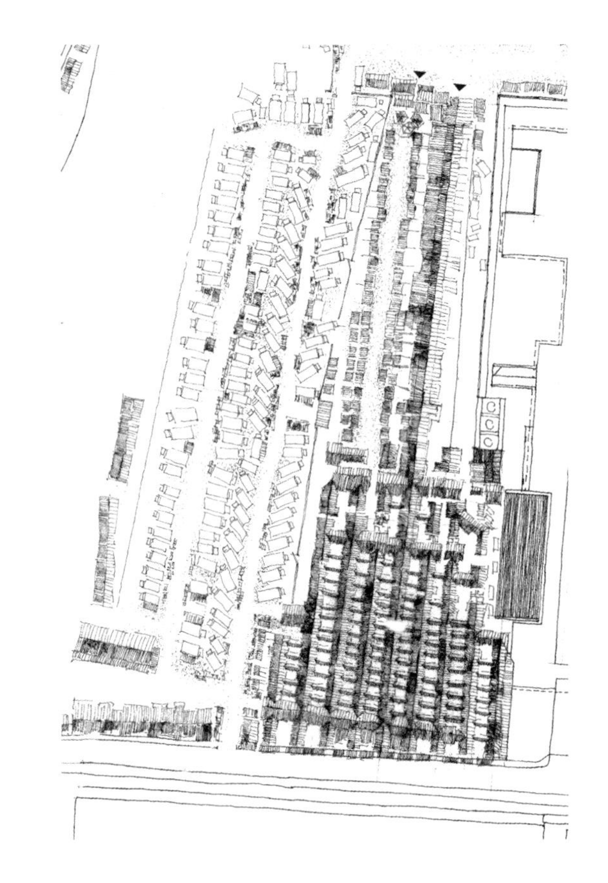

Above: Sketch plan of impromptu clothes market to the west of the burnt out sports centre
Below: Early concept plan for a pedestrian route from the Grand Hotel Pristina to the impromptu clothes market via the Community Canopy

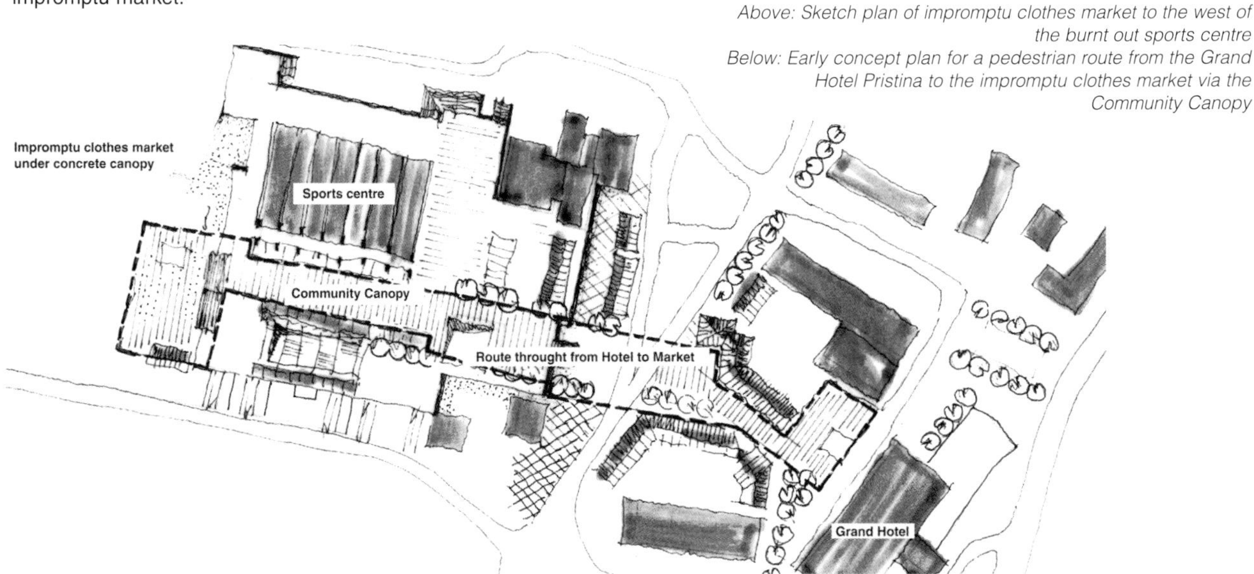

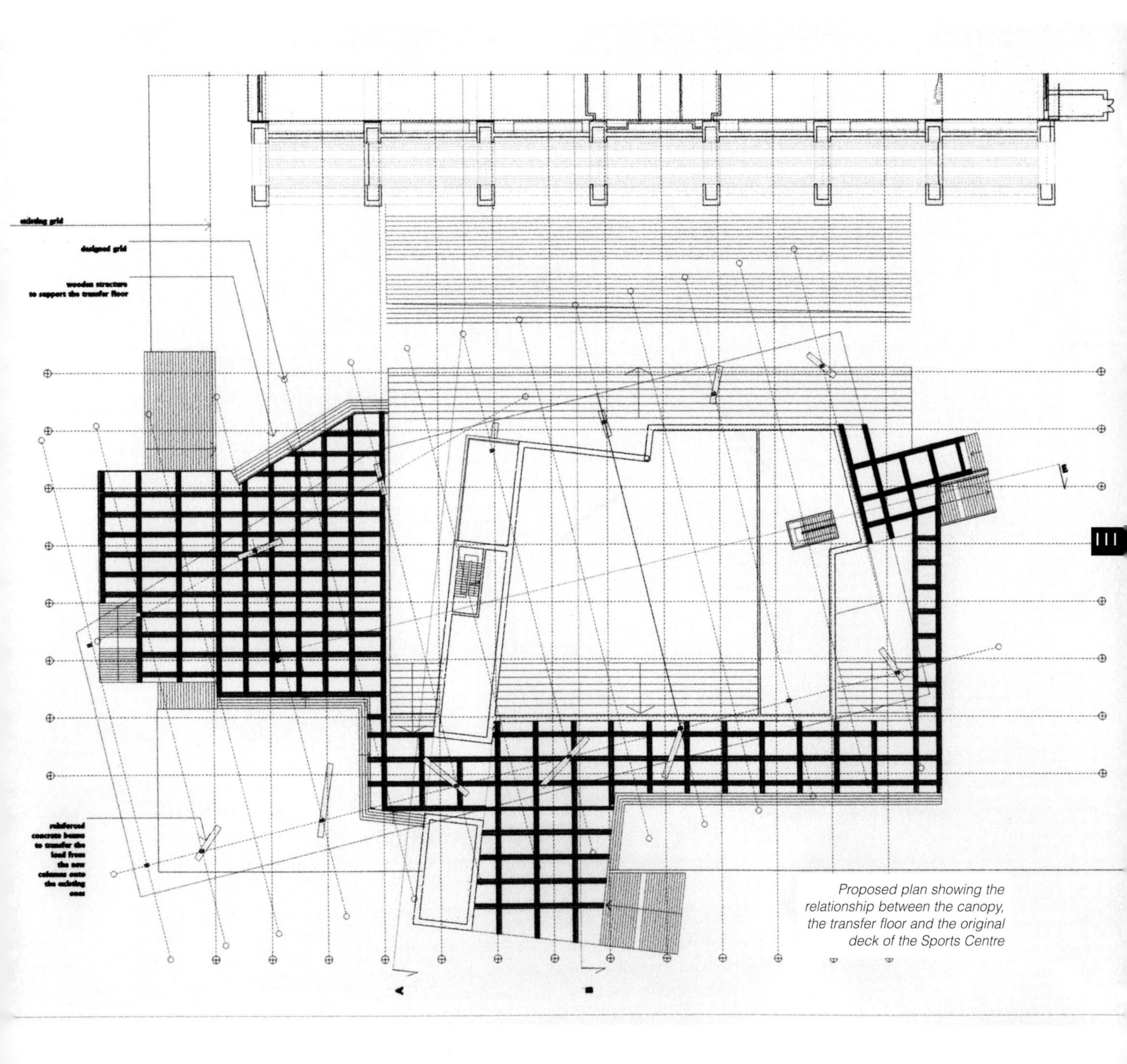

Proposed plan showing the relationship between the canopy, the transfer floor and the original deck of the Sports Centre

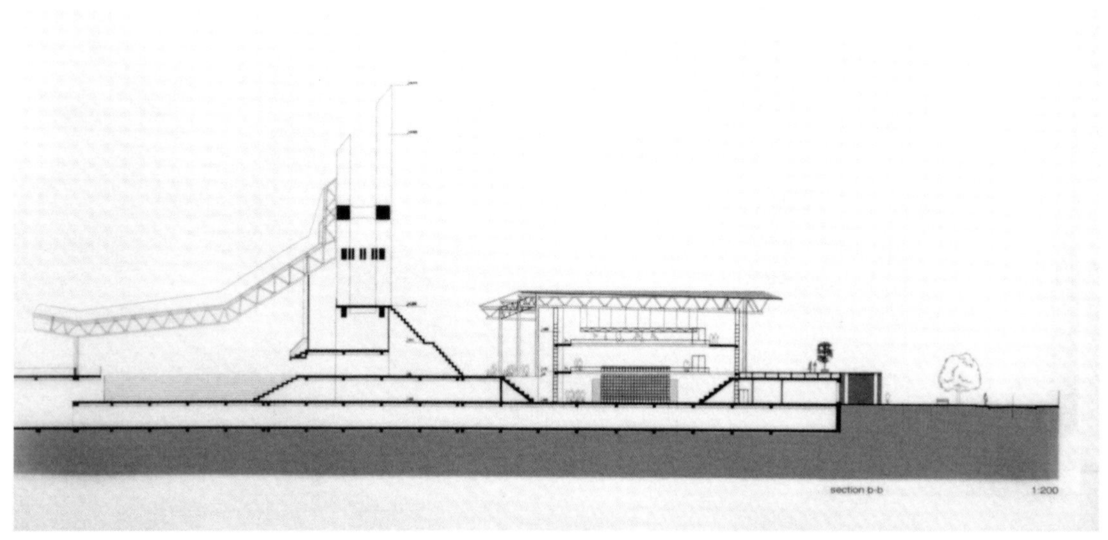

Above: Section through new theatre alongside the remaining half of the original Sports Centre roof
Below: Section through the new theatre under the community canopy

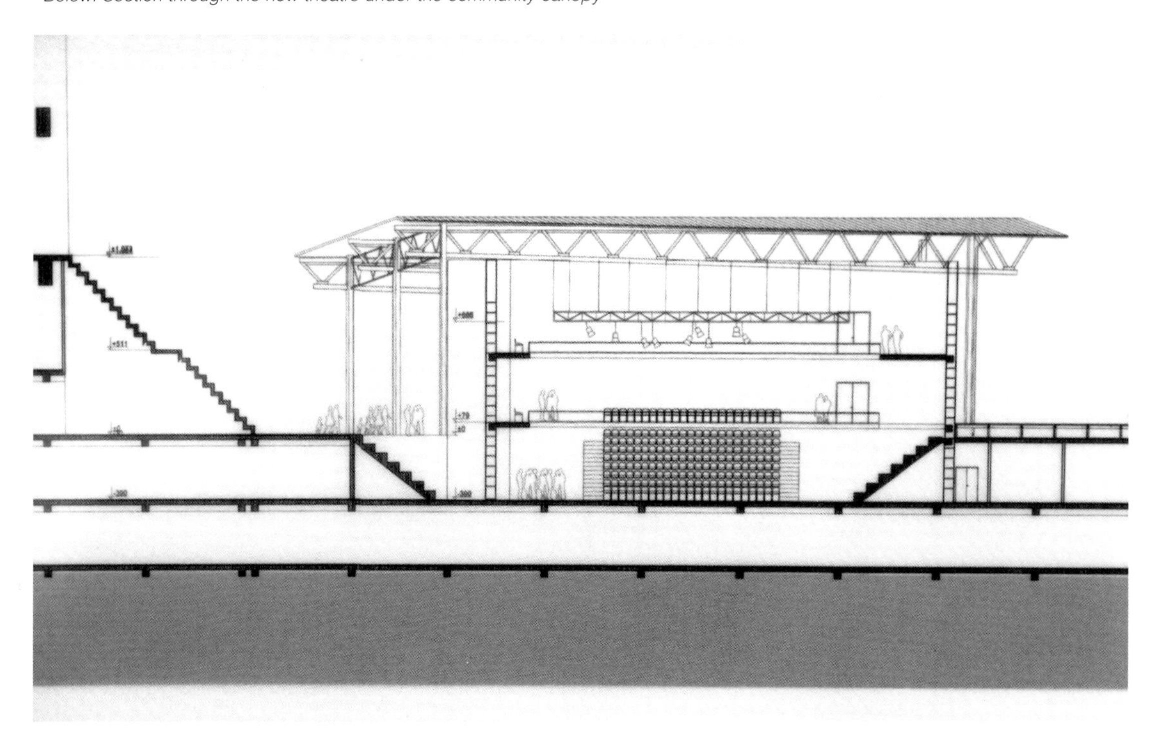

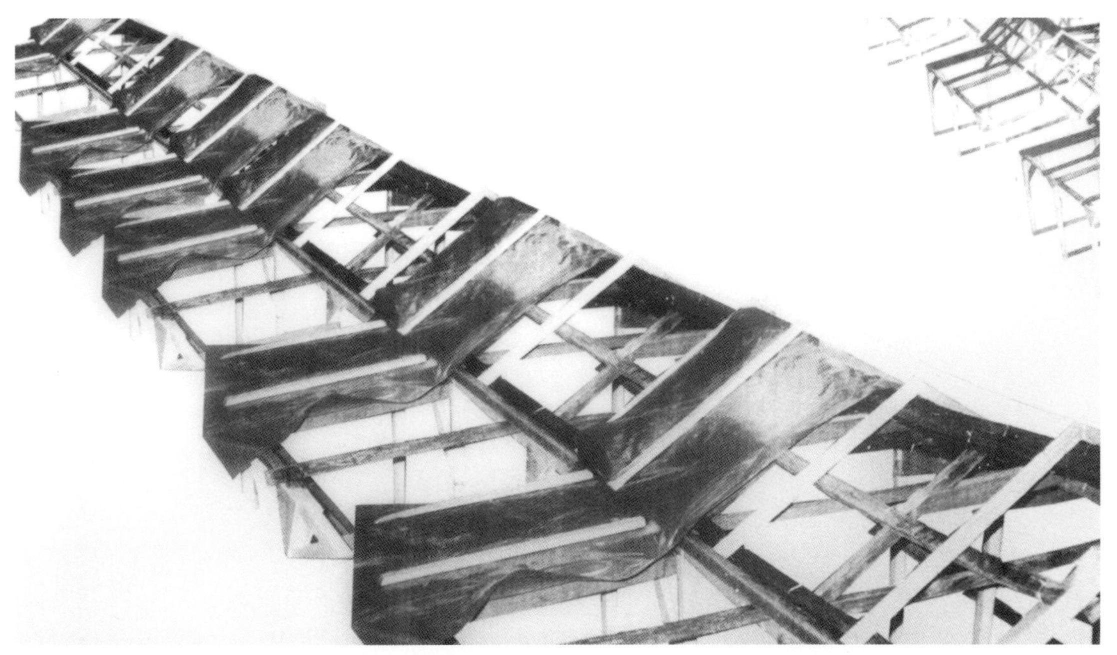

Above: Modelling experiment at CAT. Making a polythene scrap timber canopy
Below: Collaged view of the community canopy from the street

A traditional Turkish house

Chapter 5
Loose-fit Technologies

This chapter is a review of the technologies proposed in the student projects. A more detailed set of references is given in the Bibliography. Students observed at first hand the use of both traditional and contemporary building materials in Kosovo. Elements of both were chosen by each student to create a palette for each individual project. This is covered in 5.1 below.

In addition, students made full use of the opportunity to reuse old, damaged or destroyed buildings and building materials. Inspired by the landscapes they found during the field trip, the students proposed the use of building, energy-saving and generating technologies based on locally available raw materials and new 'green' or 'alternative' technologies they had come across elsewhere (either through previous study or encountered at the Centre for Alternative Technology). This is covered in 5.2 below.

A wall and door in a traditional house

5.1 Materials

This section looks at the traditional and contemporary building technologies already in use in Kosovo that the students considered adopting for their projects.

5.1.1 Traditional construction

The lower walls of the older traditional 'Turkish' houses had stone plinths surmounted with fired or sun-baked bricks, both laid in mud mortar. The walls of the upper floor were braced timber framed structures with panels infilled by wattle and daub. Suspended floors were constructed from timber joists and floorboards. Roofs were pitched timber frames clad with fired-clay 'Roman' tiles.

Walls might be mud plastered and painted with limewash. To give some protection to the inhabitants from falling masonry in the event of an earthquake, the better built properties would have timber 'ladders' inserted within the wall at regular intervals of a few courses of brickwork. Large pairs of timber doors, with stout timber frames, would give access to the property through high walls surrounding the courtyard.

5.1.2 Contemporary construction

The growth in the use of timber for building has not been sustainably managed. Even the poles cut locally and marketed extensively are a rapidly diminishing resource. Consequently, sawn structural timber is scarce and expensive. Mud brick walls are not repaired and so decay more rapidly than they used to. Hence walls and suspended floors are framed with reinforced concrete columns and beams and infilled with extruded fired-clay bricks produced in a single factory sited on the border with Macedonia. Similarly, mortars and plasters now have a cement rather than a clay binder. Despite its scarcity, timber is still used for roof framing. 'Marseille' interlocking fired-clay roofing tiles are pitched at 30 degrees to give what the students regarded as a 'Swiss Chalet' image.

A contemporary reinforced concrete framed shop house with timber roof structure during construction

Whilst sawn timber needs to be brought into the region there is a rapidly decreasing local supply of timber poles. In the past, poles were cleft to provide a ventilated cladding to upper floors but now they are primarily used for domestic fuel and scaffolding.

Above: Cleft poles used to clad the upper floor of a traditional house
Right: Trailer of poles for firewood and scaffolding

Top right: Internal decorated plasterwork
Below right: An undercoat of traditional balled mud plaster in Vushtrri
Below left: Public space in Vushtrri with walls plastered with a contemporary site applied patterning method

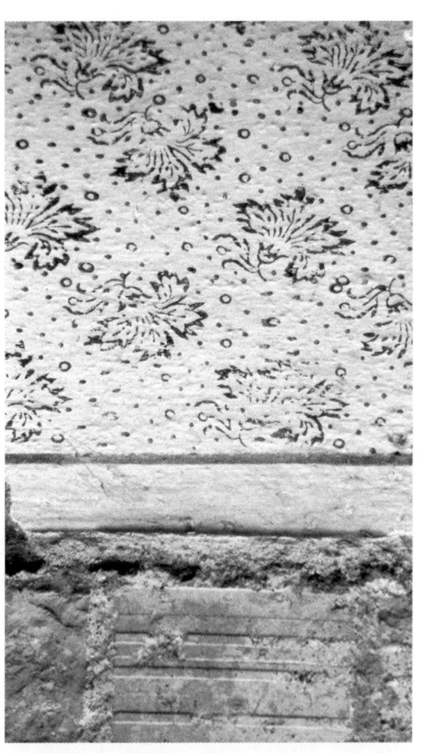

5.1.3 Student proposals

Mud bricks

Students wanting to involve children in the therapeutic and educational aspects of building production reintroduced the use of mud bricks and plaster. This was justified primarily by its hands-on immediacy allowing for individual self-expression in and on the wall surfaces.

Poles used as scaffolding in Vushtrri

Timber poles

The London students regarded local timber poles as a valuable source of flexible framing. Poles needed to be carefully sorted (according to timber species, size and durability) and then treated with preservative. In most cases a system of bolted steel shoe joints was devised to allow the poles to be used in complex framing systems for roofs (Chris Hale's portal framed swimming pool), towers (Naomi Day's diving pool) and secondary cladding structures (Domino House, Ashkali Area).

'Onduline' undersheeting

Chris Hale also imported a cheap undersheeting of corrugated, bitumen impregnated organic fibres (trade name: 'Onduline') which allowed the roof cladding of local 'Marseille' tiles to be laid at a reduced pitch of 12.5 degrees. This reduced the weight and height of the roof without which such a long span of interlocking tiles would have been unfeasible.

The chalet roof

Ben Brown's 'idea about a roof', which would provide a canopy across three adjacent burnt out concrete framed structures, came from his perception of the timber 'Swiss Chalet' hat being formally distinct from its reinforced concrete framed base. He noted the flexibility of its relatively lightweight structure using hips and valleys, ridges and overhanging eaves to 'fix and join' the three given structures below. He intended to make savings on the quantity of scarce timber used by sawing the 125mm square imported baulks into more structurally efficient 40 x 125mm rafters.

The 'FERT' system

Andrew Fortune and Steve Citrone were fascinated by the local 'FERT' system for constructing suspended floors, which they adopted in their projects. The framed reinforcement for floor beams is partly cast in concrete in a U-shaped clay tile permanent former remote from the site in two standard lengths (4.5m and 4.8m). These are light enough to be manhandled into place without a crane. Extruded, hollow fired-clay floor blocks are then placed between the beams and the concrete floor poured over the top, obviating the need for sophisticated timber formwork.

Right: A FERT roof with clay pots and reinforcement in place prior to pouring the concrete
Below: FERT beams before placement

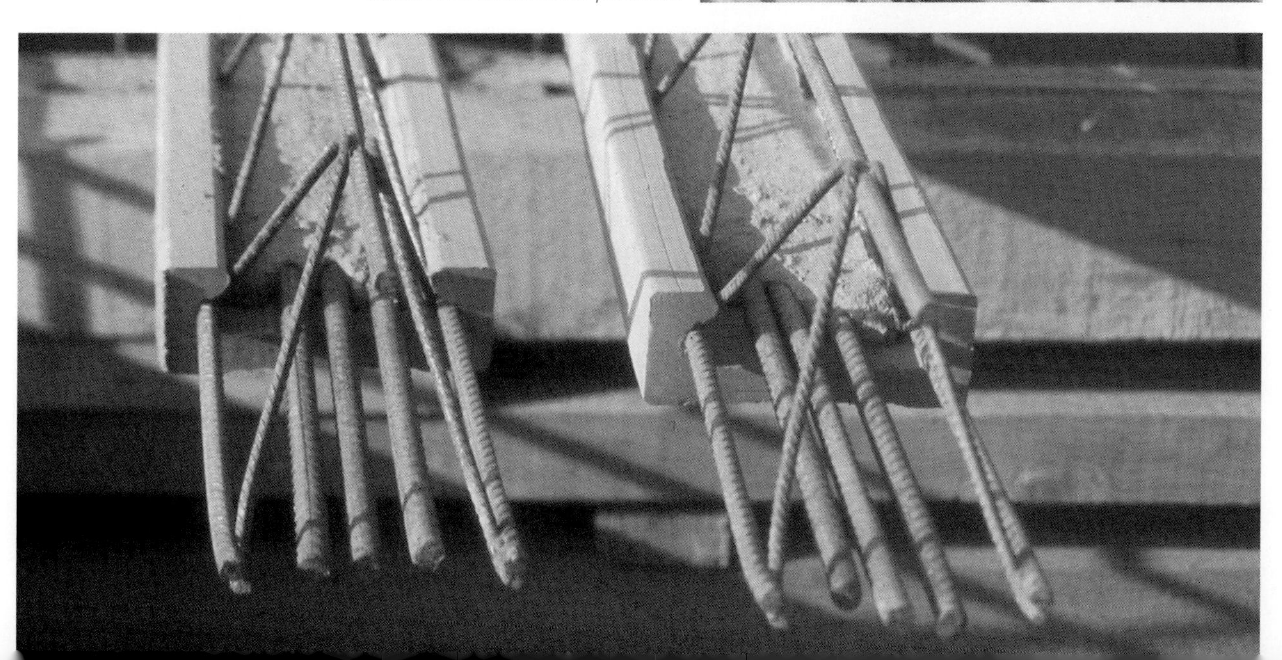

5.2 Recyclable landscapes

This section covers the reuse by students of the various types of landscape observed during the field trip. It also reviews the various 'green' or 'alternative technologies' adopted by the students within their design proposals.

The Domino house in the devastated Ashkali Area of Vushtrri; the burnt out Sports Hall in the centre of and redundant sheds on the outskirts of Pristina were all reused by the students. Rozia Adenan recycled military and industrial cast-offs (tyres, sand bags, wire and mesh) to build barricades to protect children at play. Andrew Fortune initiated the construction of a bridge made from discarded plastic bottles as a way of communicating with the inhabitants of the polluted River Site.

5.2.1 The looted Ashkali landscape

The few remaining serviceable reinforced concrete frames in the Ashkali Area needed strengthening. They had not been designed to stand independently of their brick and block infill, and repairs to the concrete and the replacement of sheer walls had to be a priority in this earthquake prone area. Several schemes over-clad the Domino House with a framework of timber pole scaffolding, buttressing the damaged structure and creating a narrow zone of access around the perimeter of the building. This allowed a complete transformation of the façade, whilst taking full advantage of the existing structural frame. Students proposing a regular recladding of the frame did so to symbolise the need for a continuous renewal of the community's commitment to a sustainable public realm. The alternative would have been to reclad the frame with a new 'permanent' skin, hiding the Domino skeleton, which was so symbolic of the tragedy of ethnic cleansing, behind an image of undisturbed timelessness.

The looted and vandalised Ashkali landscape

Crushed concrete, recycled bricks and roofing tiles

But what of the majority of collapsed buildings? At first, students imagined a huge machine crushing up the broken concrete and brick into gravel sized pebbles for recycling in new, mass concrete foundations and floor slabs. Then they noticed that neat piles of cleaned up bricks, blocks and roofing tiles had been stacked outside hastily assembled habitations within the ruins. Scavenged by squatters, these piles indicated both the stored up wealth of the occupants and the threshold they had laid down between public street or footpath and their private home.

Gabions

How could the remaining chunks of concrete rubble be similarly transformed into valuable building elements for the occupiers without machinery? Gabions, small wire cages assembled in place and then filled with broken rubble (100mm to 150mm diameter lumps), were proposed by many of the students as a way of turning masonry waste into walls with the minimum capital investment.

Allotments

After the rubble was cleaned away most students working on the Ashkali site proposed allotments placed in newly greened plots, sustaining the out of work, newly urbanised migrants in their recently secured landscape.

A damaged reinforced concrete frame reused to build a shop / house

5.2.2 Industrial/urban landscapes

Generally all new buildings in Kosovo are constructed using reinforced concrete frames. However, in the recent past some large mild steel frames and long span roofs have been fabricated abroad, imported and assembled on site. Unlike in other areas of Europe, the students did not see vast long low areas of industrial sheds. Instead these structures tended to be placed singly, stark, skeletal and often redundant against bare brown fields. The massive burnt out steel roof of the Palace of Sports was, by its size and location in the centre of Pristina, a reminder of a heroic past, its worth literally reduced to its scrap value.

Reuse of steel members in Pristina

Students working on the Roundabout and Sports Centre sites reused redundant steel structures to construct their proposals. Reuse of steel elements in their original form, with making processes restricted to just cutting and drilling, is preferable to recycling the steel from scratch as treatments such as galvanising or powder-coating can inhibit the process or cause pollution. Being already on site the reused steel members consume very little extra energy in their transformation.

Angels Lopez developed an interesting structural hierarchy of materials. Her hostel for internally displaced people used timber (locally sourced poles and sawn sections) to construct cabins and frames on the lower floor. Reinforced concrete portal frames (the current default option) for the upper floor were covered by an over-arching canopy of reused steel.

Solar panels and photovoltaic cells

The roundabout site contained a number of single-storey sheds, the largest of which was covered with a saw-toothed roof with north facing rooflights. The reverse slope, facing south and pitched at a suitable angle to catch the sun head on, provided a unique opportunity for the installation of both solar panels (producing hot water) and arrays of photovoltaic cells (producing electricity). This sunny, silent generator of heat and power contrasted with the nearby dilapidated and polluting district heating plant, which was incapable of producing an adequate electricity supply.

5.2.3 Peripheral grass landscapes

Vushtrri is the hub of a rural economy, with coarse grass hills folding right into the edges of the market town. Tractors and trailers haul firewood and vegetables over short, muddy tracks into town and take away groceries, clothes and building materials. Displaced people have settled on the often polluted edge of town where no-one else, for the moment, lays claim to the land and have begun to cultivate their own vegetable gardens. From the periphery of Pristina fingers of grassland penetrate the city, flowing around the ageing Soviet era concrete tenements and allowing free pedestrian movement. Pristina students, when discussing the design of new public space, would always prioritise such grassy parkland over hard surfacing. London students saw this landscape as both a resource for building materials (straw, earth and scrap) and as a design opportunity.

Green fingers of land lead the pedestrian from the periphery deep into the centre of Pristina

Trailers carrying poles to market in Vushtrri

Straw walls

Kosovo has hotter summers and colder winters than the UK. To provide internal comfort by mostly passive means the fabric of dwellings would ideally consist of a carefully arranged combination of thermal mass and thermal insulation. In the winter, insulation on the outside of the dwelling would trap heat generated inside within the mass of the building's walls and floors. In the summer, the time-lag provided by a thermally massive building envelope would preserve the cool of the night air inside the house throughout the day.

The students found no evidence of the use of insulating materials in construction except for the occasional newspaper stuffed into a ceiling void. A modicum of thermal mass was provided by concrete and brickwork, but the buildings were very cold in winter. Domestic heating is either electric, an inefficient form of energy when used for heating, or from open fires burning rapidly diminishing supplies of local pole timber. Electricity is provided by the opencast-mined coal driven Oblic power station, which is polluting and inefficient by western standards. This is an unsustainable situation.

Students searched around to find a suitable, locally available insulating material. In Kosovo hay stacks are a common feature of the rural landscape and some students adopted straw bale technology for their proposals, in order to conserve heat in the winter. Building with straw bales originated in the mid-western states of the USA in the nineteenth century and has recently been rediscovered and developed by self-builders in Europe and North America. Such walls are cheap and have excellent insulation value. If their details of construction are well designed straw bale walls can be long lasting. The straw can be treated with an organic pesticide such as borax. Straw bale walls can be covered with lime-based plasters or clad with various types of rain screens (corrugated iron, timber boards – as seen in many traditional agricultural buildings in Kosovo) to provide a 'breathing wall'. Breathing walls allow limited ventilation through the wall. Outgoing air passes some of its heat to the incoming air. There are no vapour barriers in the wall and, provided that the permeability of the outside is significantly greater than that of the inside of the wall, water vapour migrates to the outside of the wall and evaporates rather than saturating the straw.

Grass roofs

In a turf or grass roof, grass and local vegetation protects the waterproof membrane over which it is laid from sun, weather and traffic. These roofs increase the thermal mass of the building, useful in delaying the passage of the sun's heat to the interior during the summer. Steve Citrone's Vushtrri College of Building proposed a long flat green roof at first floor level that would be part of a public route into town. To withstand pedestrian traffic loads 'intensive' green roofs are deep and heavy and require substantial supporting structures. Similarly, Tim Wong's terraced green roofs acted as pedestrian viewing platforms. They were designed to have a low impact on the steep grassy slope of the Palagonia site, sheltering a 'hammam' (Turkish bath) whilst at the same time opening up the terraces as planted footpaths with a view. Other students proposed lighter 'extensive' pitched green roofs supporting moss and lichen growth that were not intended to carry live loads but still maintained a green cover to the landscape on the edge of the settlement.

Above right: UNMIK issue polythene tarpaulin used to patch up a damaged house for the winter
Below right: Polythene and scrap timber roof of the new impromptu clothes market in the centre of Pristina, an interior view

Polythene

Large quantities of blue 1000 gauge polythene sheeting were issued by the United Nations to help refugees make temporary repairs to their damaged houses and to provide shelter for them through the first winter after the war. Polythene wrapping – whether as a temporary cladding to the Domino House, or as a broad canopy of ridges and valleys over the temporary clothes market behind the Sports Centre in Pristina – provides a striking landscape. Polythene's ghostly, ephemeral qualities are perhaps best displayed when it is spread out to hover, glistening in the sunlight, above the undulating contours of an extensive market garden.

If polythene is to serve its purpose as a sheltering, waterproof, translucent screen it needs to be stretched, wrapped and bound around a rigid structure, held flat in rigid rectangular panels between nailed timber battens, or thrown over a tensioned net. Plastic sheeting is also available in a variety of colours, reinforced with mesh, and cleated with riveted tie eyes (as for lorry tarpaulins). Galvanised steel pipe can be used as a rigid framework over which polythene can be stretched to form polytunnels. The more ambitious long span polythene structures of James Ross's market gardens in the Ashkali Area have their origins in contemporary Spanish greenhouse structures, which cast reinforced polythene over stretched wire lines propped up with concrete or timber posts.

5.2.4 The polluted riveraine backlands

Both Vushtrri and Pristina turn their backs on their watercourses. Watersheds edge and criss-cross the settlements using only gravity to disgorge liquid effluent. A barrier to vehicular traffic, they are a short cut for children, a shelter for wildlife and a place of quiet and surprise amidst the bustle. The scale of these linear spaces changes with the seasons. Summer tunnels of flooded foliage give way, in winter, to pungent open wastelands of rubbish and litter.

Looking down over the sea of UNMIK polythene which comprises the roof of the new impromptu clothes market in Pristina

Water supply

Despite the floods and the river, a steady supply of clean drinking water is still lacking in Vushtrri. The students experienced this situation firsthand as they had access to only two hours of water supply per day during their stay. In Kosovo, the problem was less that of quantity of rainfall and more of the management of its collection, storage, treatment and distribution.

The Vushtrri River Site is an example of this problem in microcosm. A small trickle of a watercourse in summer, the river floods the surrounding houses with polluted water in winter. Inhabitants reported that they spent the first two months of spring cleaning up after the flood. Andrew Fortune proposed damming the river upstream to curb the flood, and storing the winter excess water for use during the summer drought. He designed a ring main supplying the water behind the dam to the houses in the vicinity, using an aqueduct system inspired by Alvaro Siza's scheme for Avora, Portugal. Each house would also collect rainwater from its roof, have substantial ferrocement water storage tanks and a sand filter for purifying drinking water. Surplus rainwater would run into the ring main and so into the reservoir behind the dam.

Pipes from bottles

Andrew intended that pipes used in the aqueducts would be made from recycling empty plastic bottles recovered from the site via a process, developed by Aston University, eliminating the need to melt, compound and pelletise the waste plastic prior to extrusion.

Reed beds

To avoid pollution from 'grey' and 'black' water the students proposed traditional piped sewerage systems. However, once delivered out beyond the town, this pollution was converted into plant nutrients via an array of reed beds. Chris Hale laid out his beds of waving plant life as broad horizontally fed meadows, interspersed with allotments. Tim Wong and Lole Mate proposed vertically fed terraced parkland on the steep hillsides of Palagonia, connected by gently sloping footpaths and stairways leading to the cleaned up and planted valley bottom. Organic waste recycling on this large scale would transform the landscape.

District power and heating

The orientation and steep inclination of the site suggested to Tim Wong and Lole Mate the potential for an optimistic landscape of communal pollution-free endeavour. They modelled the heat and power generation for their Palagonia schemes on the example of district heating used at Lyckebo, Sweden, where 550 dwellings are powered with extensive arrays of PV and solar panels.

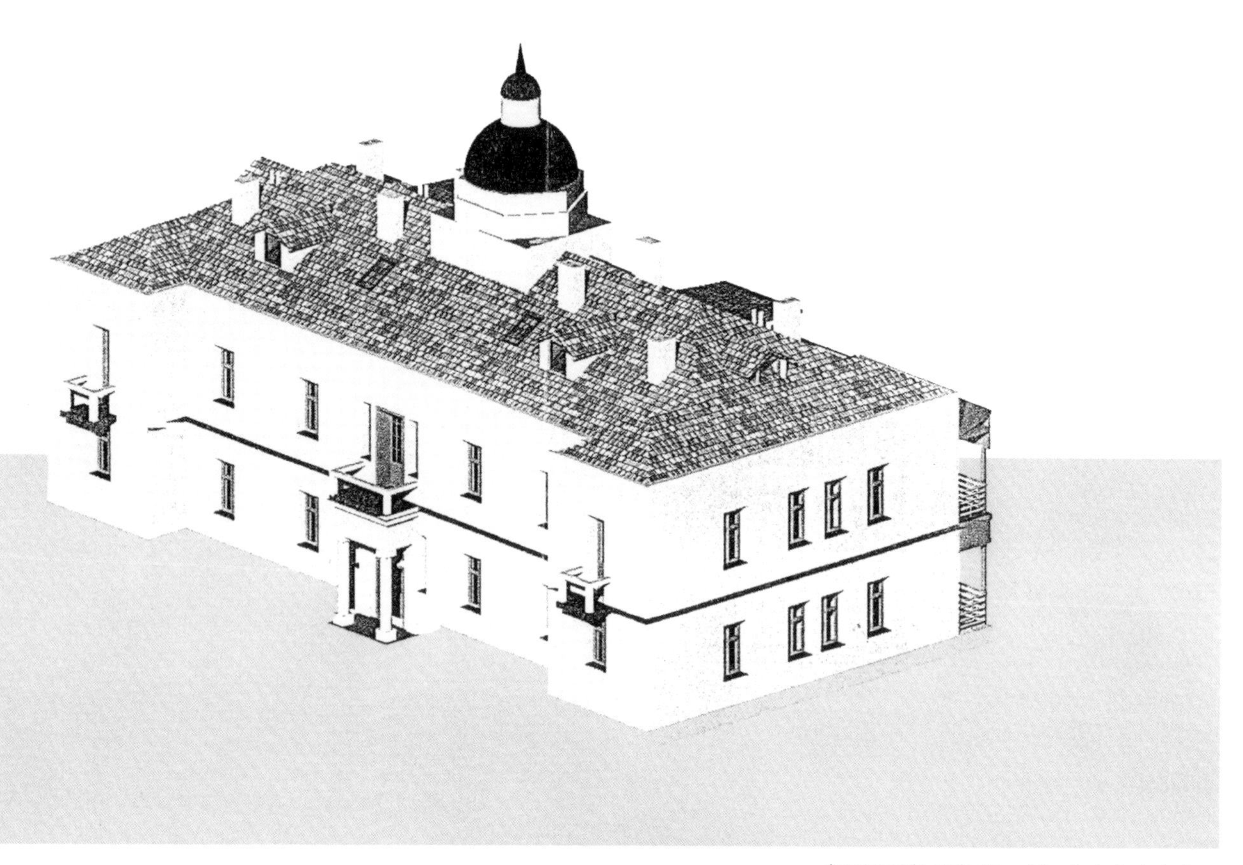

Axonometric projection of the Ambulante before conversion

Chapter 6
Lessons for the future

This last section draws on a conversation with architects Tim Booth (TB) and Helen Little (HL) who were our hosts in Kosovo for the last year's field trip and from a report by Professor Lynn Davies on the British Council's support to Pristina University. Helen and Tim were running Development Workshop's office in Vushtrri and their remit was to re-build, over two years, a large number of houses which had been damaged or destroyed in the war. The chapter reflects on some of the lessons that the Kosovo programme has had for teaching, practice, and the encouragement of local initiative in situations of rapid change. Taped quotations and extracts from the report are in italics.

Loose-fit construction in an unfamiliar situation

One of the intentions of the teaching methods was to broaden the approach of students, to move away from focusing only on the design of finished products, and instead to concentrate on the way a variety of alternative appropriate outcomes might be derived from the materials and skills available.

In an unfamiliar development situation it was easier for students to develop an overview of processes relating to everything from macro funding concerns to detailed formwork design. Having exposed the students to a different social and technical context, we found them to be in a better position to question the way architects work in the UK.

What we tend to design in the UK is a finished product. We cut through it in various places to show how the assembly works but we never design the way it is assembled. But (in Kosovo) we were trying to work out how to minimise the amount of material used, using the products given to us. We spent a lot more time designing how you do things than designing buildings as a product. Designing process in a looser sense of the term. (TB)

In applying this 'loose-fit' approach to the process of making buildings, students used traditional skills and materials and recycled landscapes, together with imported products and ideas, releasing the dormant potential of unused local materials. They assembled their proposals in a spirit of experiment, looking for ways in which each new piece of the jigsaw could fit easily (and loosely) with neighbouring building elements (present and future).

In the UK, if a 'product' (social or technical) presents a problem it is replaced by another 'product'. These 'products' are standardised and developed remotely for a wider market. Consequently, they tend to have a low tolerance for their neighbours and respond badly to unforeseen problems. A loose-fit approach tackles problems locally by drawing in ideas, adapting precedent and developing prototypes in the situation within which they are embedded.

Extending the use of local pole timber, beyond that of scaffolding, to provide a developed long span portal frame structure – and the use of imported 'Onduline' sheeting to facilitate the adoption of local clay tiles to cover such structures (Chris Hale) – is just one example of this approach.

Becoming familiar, gaining confidence and broadening access by using a restricted palette of skills and materials

The range of building materials available in the UK can be bewildering to a student and in many studio projects students are lost as to which materials to use. In Kosovo materials and skills are both restricted and very visible, as most work is carried out on site rather than in a factory. Thus it is easier, in Kosovo, for students to discover ways to build their projects. (In fact one of the ways to produce beautiful structures is to restrict your palette. Old cathedrals made only of stone, and Le Corbusier's works of just concrete, for example.) Within a limited kit of parts it is easier to work out how to make things and this inspires confidence.

In Kosovo you are very limited. You have some concrete, which is expensive because of the shuttering, but this is reused. You have fired hollow clay blocks, plaster, a limited range of sanitary ware and a wood burning stove. Roof structures are timber (covered with clay tiles) and suspended floors and ceilings are timber boarded. We are not talking about a theoretical language of modernism here... (HL) Methods of building are completely different to those used in the UK, although when you see a properly constructed building it looks as if it could have been made in northern Europe. You have to build a fairly complex traditional timber roof on site using only a hammer and an adze and probably a chain saw. No screws. (TB) You buy your builder a shirt and you hang it from the roof structure, in its packaging or not, before you clad the roof. If you can't afford a new shirt you see a towel...a towel is the symbol of the Kosovans. (HL)

The process of building houses is embedded within Kosovan culture. The skills required are limited and families get fully involved in constructing their own homes. As a result, changes, repairs and additions to domestic construction are relatively easily carried out – with minimal permissions or importation of specialist skills and materials required. Such simple and transparent building methods not only made it easier for the UK students to understand and represent the process, but also broadened the inhabitants' access to the creative and therapeutic process of manipulating their own physical environment.

In their proposals for a new public realm students aimed to involve users in the phased construction and continuing adaptation of their schemes. For example, Yenny Gunawan and Phaidon Perrakis proposed that the recladding of the Domino House be a moveable feast, carried out as a regular bi-annual children's workshop. Similarly, Magda Raczkowska proposed harnessing the energy of the local traders, who had already constructed an extensive polythene covered clothes market, to facilitate the occupation of the burnt out shell of the Sports Centre.

Development tourism?

Time to listen, model and articulate new ideas

It is only when you stop, observe and reflect, with no agenda other than to look, listen and learn, that really open communication can begin: conversations can be exploratory and local individual creativity can be encouraged within the community.

We are all going to remember [Andrew Fortune's] water bottle bridge, which was just magical. [This actually constructed, instant, on-site project] was based on real needs and problems. It was optimistic and the great thing about it was its engagement with local people. (HL)

The best bits of our job were when we had positive engagement with the local people. But this was very difficult because buildings were produced very quickly without any long drawn out bureaucratic process. It was really fast track. You would shake a hand [of a beneficiary] once at the beginning and then try to remember their name when you saw them again at the end. (TB)

The students spent a lot of time with the villagers, not just elected representatives, but those who were there on the street, who were there playing. They had a healthy interaction at that level. There was no stronger agenda and they could mix and talk regardless of language. They were not promising anything and did not have the greedy running up to them. You just got the inquisitive, which produced very interesting results. The students did not have the strict programme which we had and which did not allow us to have this type of productive interaction. (HL)

In classic development projects, the framing of proposals usually comes from outside: from NGOs, from planners, developers and government. Local individual creativity, which can often be the key to a successful project, is seldom noticed by these outsiders. In contrast, because of their close and intense, if brief, involvement with the local situation, some students were able to give a voice to local initiative... Rozia Adenan extended the idea of protected play spaces for children from a prototype she had observed during her survey. Liz Crisp developed a metaphor for the transformation of damaged building shells into temporary public buildings from her recording of the use of the veranda of a nearby house by a squatter family. James Ross used his observations surrounding a number of chess matches, held in the most unlikely places, to transform the spaces in his scheme.

Students were also able to make proposals for the Ashkali Area, where reconciliation seemed so out of the question that no-one had thought it worth developing detailed proposals. Development Workshop had found it impossible to rebuild in this area because of the animosity left over from the war. Yet the students were able to produce proposals that had, at least, highlighted possibilities for discussion within the local population.

The UNHCR were very keen to assist the minorities. [They] were desperate to get people back. We weren't able to interview the Ashkalis directly, as they lived in Novi Sad in Serbia. We had to commit to assist them and organise a 'come and see' visit. We had a very intense three days with a very high level of security. They still felt very vulnerable. They had lived away for 2 or 3 years and had managed to build up something in Serbia and they didn't want to return unless enough of them were going to return. It became very unworkable.(HL)

Hand grenades thrown over the back walls of Ashkalis living in Vushtrri was a fairly regular occurrence. The Ashkali families still living in Vushtrri were afraid that if more Ashkalis started to come back then the attacks would become worse. (TB)

Proposals for the Ashkali Area were presented to the UN Municipal Administration, and the newly elected mayor, in Vushtrri in the form of an exhibition and seminar by students from both London and Pristina during the second year's field trip. These proposals were intended as a contribution towards a dialogue that might end the uncertainty surrounding the Ashkali Area and might, in turn, promote its rebirth.

The Green agenda

Students were encouraged to make locally sustainable proposals: proposals that would have a significant impact on improving the environment at a low capital cost. Many of them proposed improvements in metal fastenings for timber structures and the recycling of metal structures, where available. Insulation, alternative energy generation (Lole Mate and Tim Wong), sustainable water supply (Andrew Fortune) and reed bed treatment of waste water and sewage (Chris Hale and Keith Smith) were included in many student schemes. Working in this way the students provoked debate on these issues which were, whilst essential for sustainable development, not directly relevant to the emergency relief effort (since funding was for houses to be rebuilt as they had been before the war).

Tim and Helen brought up two simple opportunities for improving sustainability, which could reasonably easily and effectively have been included in their programme.

On saving timber:

There was an idea that they could slim down the timber sizes they were using by 50 per cent. You could cut the 10cm section in half even allowing for the waste. They could do that in Kosovo because there were sawmills and then use gang nailing with plates to make simple timber trusses and erect roofs very quickly. But then you are up against the construction industry which has a way of doing things plus a social problem. [Kosovans] used the roof space for storage of winter food and they were suspicious of anything which changed the vernacular way of dealing with things. (TB)

On insulation:

They didn't like our lintels. They take 3 or 4 bits of reinforcement, a piece of board, lay them across on the board, splat some concrete in and put blocks on top. The lintel itself would be the depth of the reinforcement bar and you could usually see them. Structurally you think that this is insane. It wasn't failure to understand a structural problem. They realised that a lintel formed an enormous cold bridge compared with all the hollow clay blocks, so they got condensation and mould forming. People didn't want lintels because they had no insulation. When you investigate methods of building you find that they are far more difficult to unpick and to generate new solutions than you might first think. (TB)

The reason for non-inclusion of these sustainable development improvements was the nature of the relief effort itself.

The Green agenda and emergency relief

The green agenda was very valid. The Oblic power station worked off hideous amounts of coal even though it only worked for 12 hours a day. [Because our work was emergency relief] we were not responding to environmental issues, not using solar panels to heat water, or [even introducing] insulation for instance. We were [also] not trying to produce [building materials] locally. There are a lot of local resources…local aggregate for example. We would like to have seen this sort of thing happening within the work we were doing…(HL)

It was in the context of conversations with the Pristina students that the debate over the 'green' agenda listed above had the most impact. Debate at the Pristina School of Architecture was focused on balancing ideas of architecture rooted in a nostalgic view of the Kosovan vernacular with those of a western 'High Tech' architecture that relied on quantities of steel and glass unheard of in Kosovo for its realisation. The idea that you might produce a modern architecture using sustainable local materials, services and energy supplies, in the Kosovan context, was relatively new.

Occupants rethinking their ambitions

When change is rapid and the resources dry up, taking stock of your position, your aims and any change in the direction of your ambitions must be given time. Once new power structures have been established, however, and new ways have become customary, the opportunities for fundamental rethinking of change are reduced. Students were encouraged to explore this window of opportunity to make original and optimistic proposals.

The possibility for involving architectural students in how we might intervene is relevant here because it is in the emergency phase that you can instigate change more easily.(TB)

Jean Dumas's scheme takes advantage of this window by urging the squatter inhabitants of the 'Archigram' block to seize control of a nearby empty warehouse, for use as cinema and workshops, before the new government took it back. Naomi Day seizes the future with both hands with her optimistic entry of a diving team into a future Oylmpics. She has the team demonstrating their skills to their fellow citizens from a diving tower in the Bus Stop Square of Vushtrri. Namee Im's scheme proposes the long term empowerment of the drivers of horse-drawn taxis by modelling their phased investment in a hostel and workshop scheme.

Civil society and public space

During the war people hid inside their homes, afraid to come out, leaving the streets and other unprotected space a largely uninhabited wasteland of fear. Before the war most Albanians were not involved in public life and the public places of significance were different for each community.

[Attempting to create a new civil society was not naïve but] if you hold people's hands for so long…The Serbs had run the public sector and the Albanians the private sector. The Albanians had had very little responsibility and very little training. They were disempowered: not used to holding any [public positions]. The Serbian structure was then replaced by the UN and by NGOs. [We were] not empowering people to take responsibility for their own lives and for their own country. (HL and TB)

Most of the student projects involved the creation or rediscovery of public places. These included open urban spaces such as the public demonstration route, which Woo Song devised, and the Castle Square project by Ray Leung, which recognised and enlarged upon the current occupation, by young people, of a space between an old castle and a squatted block on the main street of Vushtrri.

Development Workshop's role in building in the public realm

The students' recognition of the need for constructive development in the public realm was recognised by the team at DW. Despite there being virtually no money for public spaces and buildings, and because of their architectural training, Tim and Helen had been asked by the Municipality to become involved with a number of public projects.

There was no money or profit for us doing this. We were seen as professional and efficient and [as knowing] how to run jobs and make things happen, to draw them and to price them and to manage and organise them. Although our remit was only for the housing, we were interested in doing more for the community and for our own sanity. (HL)

In this interstitial role they were responsible for the design of the drainage and resurfacing of what the students had named the Bus Stop Area, for the conversion of an old health building into social housing and for a new fire station.

There was a disused bus station in Vushtrri. A concrete frame, very suitable for fire engines and the UN had set up a fire fighting force. They did actually have the fire engines and so we were employed to rehabilitate this disused bus station to be the hub of the region's fire service and as a fire training centre. It was to have a practice tower to a very high specification. (HL)

The second public project was the old 'Ambulante', a beautiful brick building with a brick dome and a big roof space. It was semi-officially being squatted. A lot of the people in the 'Ambulante' had

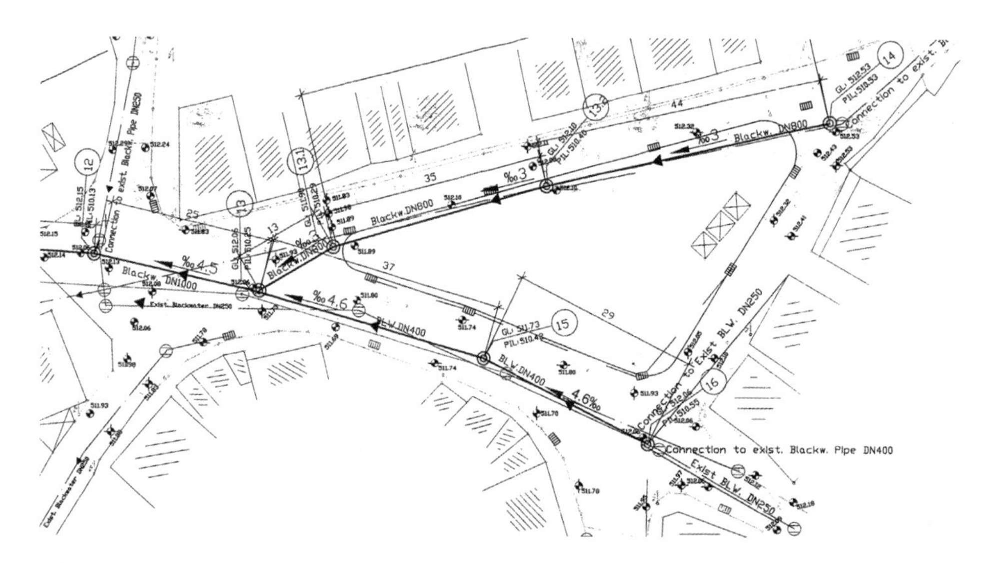

The Development Workshop/UNMIK scheme for the drainage and paving of the Bus Stop area

The doomed Ambulante, home for displaced people before conversion

House owner at work on the construction of his new house under the scheme administered by Development Workshop

The Development Workshop/UNMIK scheme for the drainage and sanitation of the River site

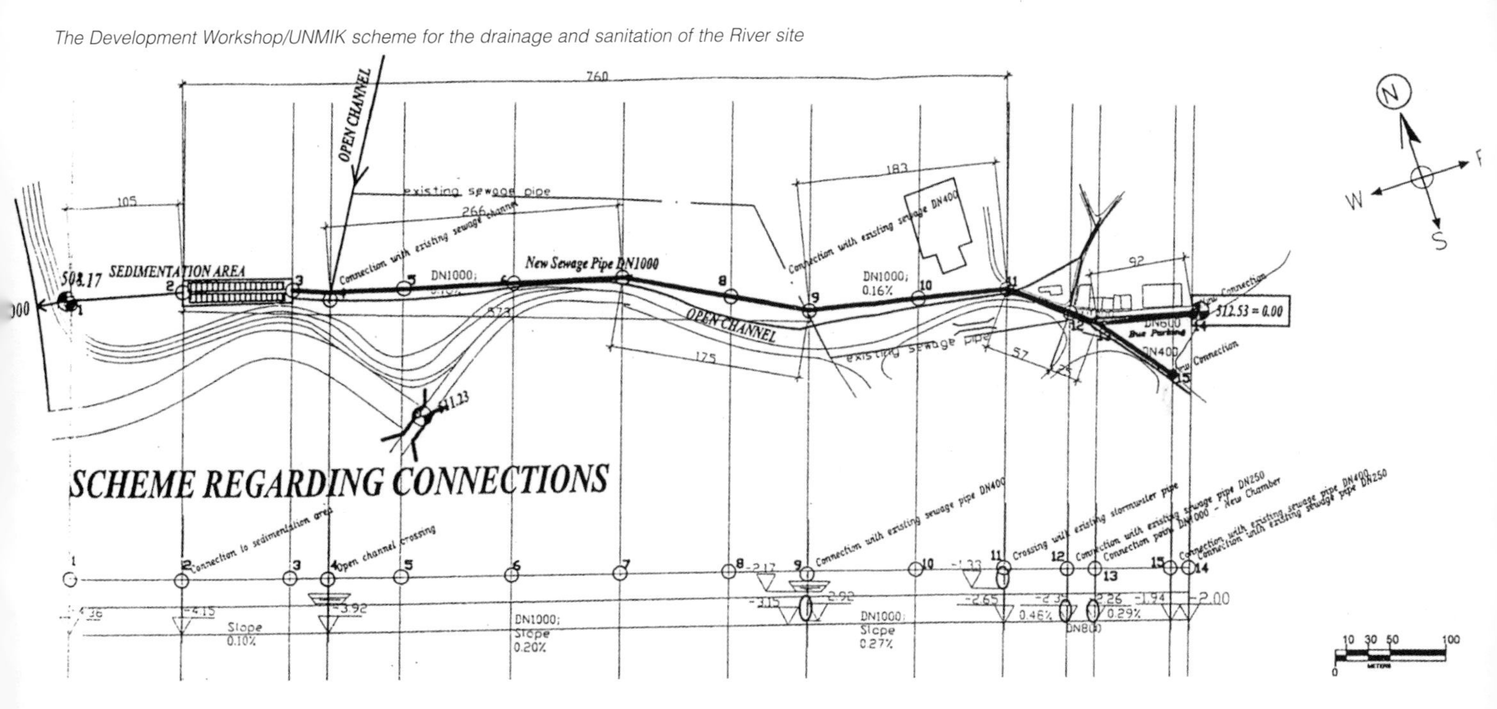

Clockwise from top left: DW beneficiary Sefedin Morina's house under construction; DW beneficiary Sejdi F. Shala's house under construction; Kosovans living in a tent whilst building their house, House under construction under the scheme administered by Development Workshop

been there since long before the war. Their houses had been destroyed through the flooding of the river. [Our job was to restore the fabric, make reasonable service provision and, by using] the roof space, to divide up the building much more efficiently. (TB and HL)

The scale of work undertaken by the students

Studio programmes and teaching methods have prioritised working outwards from fragments of closely observed daily life and details of building construction. This approach is intended to avoid reliance on pre-existing and familiar structures of thinking and maximise the take up and expression of keenly felt, empirically collected qualities within student proposals: a celebration of the unfamiliar. It is also a development of the idea that the fragment can contain the seeds of the whole (Vidler, 1999) and that therefore a full scheme can be developed from studying the kernel of truth held within a tightly bounded situation. However, Tim and Helen pointed out that this avoidance of macro-level spatial and institutional policy planning leaves many gaps in an architectural agenda for change given a situation of scarce resources.

You need to do both at once. (TB) Master planning is the missing agenda. The French decided they would put a school here, there and everywhere and KFOR decided the same. Often there would be two schools next to each other. Not just saying...here is a social space I want to put a pool in it...but looking at master planning it across the country or public spaces or theatres and entertainment. [Architecture also] has a role in structuring health provision. Do you have lots of little health centres or do you have one big hospital. How is their outreach programme? Everyone is building on the greenfield because its easier than building on the brownfield. (HL) The form of the town is being destroyed and the infrastructure that the town relies upon won't support the expansion. (TB)

Whilst student proposals were developed directly from observation and interaction with a very local situation, many of these proposals could be used as prototypes for the development of similar neighbouring situations. James Ross's sustainable market gardening community could be used as a model for other 'Gap' spaces along the main road in Vushtrri. Lole Mate's community energy plant producing solar heating and power at community level takes a community project in Sweden as its example, and might in turn be used as a reference for other south facing slopes on the edge of European towns.

Working with Pristina School of Architecture

Often a student can learn more from his or her peers than from teaching staff. Students and professionals working in the same field but coming from different cultural backgrounds often have the most to contribute to each others' education.

Whilst the students established good, albeit brief, relationships with local children and families, who were curious about and interested in their work (and for whom the results of their work were intended) it was with the local professionals, and more particularly with the students and staff at Pristina School of Architecture, that new thoughts were discussed and ideas generated. There was a real attempt at developing a shared professional discourse. This discourse continued throughout the return visit of Pristina students to London and in email contacts between students of both universities continuing to this day.

In her report to the British Council, Professor Lynn Davies comments on this dialogue:

Students had got into 'small fights' – very amicable, but relating to the content of the curriculum and the degree of a 'high tech' approach.

...the Kosovo students had a tradition of learning architecture through engineering and mathematics, whilst the UK students focused on social and urban aspects.

Provoked by the interaction, Pristina students would raise and exchange views on questions that UK students usually take for granted such as, *'whether 'Art' was a good part of architecture.'*

After working jointly on design projects, *'The Kosovo students do presentations, critiquing each others work. This is all new. They are taking photographs and measuring in the field, rather than [just] working in isolation and drawing at home. Conversely, the UK students learned how buildings are constructed in Kosovo traditionally – Kosovo students were amazed that 'some [UK] students had never done a building project!'*

Shirin Homann-Saadat, a UK student taking part in the second year of study, and who has contributed a chapter to this book, summed up, at the time (January 2002), the feelings of most (UK) students about their dialogue with Pristina students, and the experience of working in Kosovo on their lives, as follows:

Kosovo...has changed my outlook on architecture in particular and education in general. Since we started working on our projects in Kosovo I have realised that architecture really could be a tool to make the world a tiny bit better...

'Education' since Kosovo has turned into a reciprocal process. I always believed in education being a 'two-way process', but after spending years in the British education (and class) system I had started to develop some doubts and wondered if the 'hierarchical attitude' in education would 'win' (and with that destroy 'proper insights'). But since our work in Kosovo I have started believing again that education is fun, important and will leave its traces.

Loose-fit architecture

There is a need for the study of the type of 'loose-fit' architectural solutions proposed by the students and illustrated in this book. In order that the individual and collective inspiration of the occupants can more easily be explored, building elements need to fit easily and loosely with one another and with the existing landscape. They should be capable of repair, renewal or change without adjoining elements being made redundant. Instead of ignoring the effect of time, we should explore its exigencies, and make changes to old customs in a curing, healing way. In most cases radical surgery to the physical environment is not required. Proposals with a low impact on the immediate environment will be sufficient, at least until more resources have been accumulated and confidence in the future has been restored.

Even when a massive relief effort is necessary because of war, famine or disease, any emergency input should take into account what materials, skills and technologies are already in place. External inputs should seek to maximise the opportunity for the generation of confidence and creative initiative within the host community.

At a time when civil society is being re-formed in Kosovo, these student projects model ideas developed out of interactions with a society traumatised by war. As a contribution to the debate on the construction of a new public realm in Kosovo, they are not just illustrations of potential solutions but, more importantly, show ways of generating appropriate and sustainable proposals out of the particular situation.

Bibliography

General Theory and Concept

Brand, S. (1994) *How Buildings Learn, What Happens After They're Built,* Phoenix Illustrated, London.

Crawford, M. and Kaliski, K. (1999) *Everyday Urbanism,* Monacelli Press.

Leach, Neil (ed.) (1997) *Rethinking Architecture, A Reader in Cultural Theory,* Routledge, London and New York.

Rowe P. and Sarkis, H. (ed.) (1998) *Projecting Beirut, Episodes in the Construction and Reconstruction of a Modern City,* Prestel-Verlag, Munich.

Sennett, R. (1994) *Flesh and Stone, The Body and the City in Western Civilization,* Faber and Faber, London and Boston.

Vidler, A. (1992) *The Architectural Uncanny,* The MIT Press, Cambridge, Massachusetts and London.

Vidler, A. (2000) *Warped Space, Art, Architecture and Anxiety in Modern Culture,* The MIT Press, Cambridge, Massachusetts and London.

Kosovo Specific

Campbell, G. (2000) *The Road to Kosovo, A Balkan Diary,* Westview Press, Colorado and London.

Davies, L. (2001) *DFEE Support to the University of Pristina, January 8th – 14th 2001, Report on Inception Visit,* Report by Senior Advisor to the British Council, Centre for International Education and Research.

Davies, L. (2002) *DFEE Support to the University of Pristina, January – November 2001, Project Evaluation Report,* Report by Senior Advisor to the British Council, Centre for International Education and Research.

Doli, F. (2001) *Traditional Popular Architecture of Kosova,* exhibition catalogue.

Nixha, S. (2001) *Kulla – Tower House in Dukagjini Region, Kosova,* Masters thesis.

Raufaste, A. (2001) *Architectural and Urban Heritage of Prizren, Kosovo,* European Agency for Reconstruction.

Cultures of Making

(includes architectural education using live studio projects)

Alexander, C. et al (1977) *A Pattern Language,* OUP, New York.

Alexander, C. et al (1979) *The Timeless Way of Building,* OUP, New York.

Carpenter, W. (1997) *Learning by Building, Design and Construction in Architectural Education,* Van Nostrand Reinhold, New York.

Dean, A. O. and Hursley, T. (2002) *Rural Studio,* Princeton Architectural Press, New York.

Levi, P. (1987) *The Wrench,* Abacus, London.

Mitchell, M. (1998) *The Lemonade Stand, Exploring the unfamiliar by building large-scale models,* CAT Publications, Machynlleth.

Mitchell, M. (1995) 'Educating through Building' in Hamdi, N. (ed.) *Educating for Real: The Training of Professionals for Development Work,* Intermediate Technology Press, London.

Pickering, A. (1995) *The Mangle of Practice: time agency and science,* University of Chicago Press, Chicago and London.

Cultures of Situation

Andreotti, L. and Costa, X. (eds.) (1996) *Theory of the Derive and other situationist writings on the city,* Museu D'Art Contemporani, Barcelona.

Deakin, R. (2000) *Waterlog: A Swimmer's Journey through Britain,* Vintage, London.

Hill, J. (ed.) (1998) *Occupying Architecture: Between the Architect and the User,* Routledge, London.

Tschumi, B. (1994) *The Manhattan Transcripts,* Academy Editions, London.

Vernacular Architecture

Mitchell, M. and Bevan, A. (1992) *Culture, Cash and Housing,* ITP, London.

Oliver, P. (ed.) (1997) *Encyclopaedia of Vernacular Architecture of the World,* in 3 vols, Cambridge University Press, Cambridge.

Oliver, P. (1987) *Dwellings: The house across the World,* Phaidon, Oxford.

Rapoport, A. (1969) *House, Form and Culture,* Prentice-Hall, Englewood Cliffs, N.J.

The Public Realm

(Rights Cultures)

Harvey, D. (1996) *Justice, Nature and the Geography of Difference,* Blackwell, Oxford.

Sennett, R. (1974) *The Fall of Public Man,* Norton, New York and London.

Sennett, R. (1996) *The Uses of Disorder: Personal Identity and City Life,* Faber and Faber, London.

Appropriate Technology

Solar Energy
Solar Water Heating
Tapping the Sun
Solar Electricity: A Resource Guide
Environmental Building
Sanitation: A Resource Guide
Constructed Reedbeds Tipsheet
(All) CAT Publications, Machynlleth.

Schumacher, E. F. (1973) *Small is Beautiful,* Sphere, London.

Spence, R.J.S., Cook, D.J. (1983) *Building Materials in Developing Countries,* Wiley, Chichester.

Stulz, R. (1981) *Appropriate Building Materials,* SKAT and ITP, St. Gall and London.

Vale, B., Vale, R. (1996) *Green Architecture: Design for a sustainable future,* Thames and Hudson, London.

Willoughby, K.W. (1990) *Technology Choice: A Critique of the Appropriate Technology Movement,*
Intermediate Technology Publications, London.

Earth Construction

Houben, H., Guillard, H. (1994) *Earth Construction: a comprehensive guide,* ITP, London.

Norton, J. (2nd edition,1997) *Building with Earth: a handbook,* ITP, London.

Wojciechowska, P. (2001) *Building with Earth: a Guide to Flexible-Form Earthbag Construction,*
Chelsea Green Publishing Company, White River Junction.

Lime Plaster

Holmes, S. and Wingate, M. (2002) *Building with Lime,* ITP, London.

Schofield, J. (1998) *Lime in Building – a Practical Guide,* Black Dog Press, UK.

Straw Bale Construction

Jones, B. (2002) *Building with Straw Bales, A practical guide for the UK and Ireland,* Green Books, UK.

Steen, A.S. et al (1994) *The Straw Bale House,* Chelsea Green, White River Junction.

Plastic Sheeting

Howard, J., Spice, R. (1989) *Plastic Sheeting – Its use for emergency shelter and other purposes,*
An Oxfam technical Guide, Oxfam, Oxford.

Pole Construction

Huybers, P. (1984) *Manual on the Delft Wire Lacing Tool,* Delft University of Technology, Delft, Holland.

Jayanetti, L. (1990) *Timber Pole Construction,* ITP, London.

Wolfe, R. (1980) *Low Cost Pole Building Construction,* Garden Way Publishing.

Ferrocement Construction

Evans, B. (1986) *Understanding Natural Fibre Concrete,* ITP, London.

Hasse, R. (1989) *Rainwater Reservoirs above Ground Structures for Roof Catchment,* GATE, Eschborn, Germany.

International Labour Office (1992) *Fibre and micro-concrete roofing tiles,* ILO, Geneva.

Watt, S. B. (1978) *Ferrocement Water Tanks and their construction,* ITP, London.

Reed Beds and Alternative Sanitation

Cooper, et al (1996) *Reedbeds and Constructed Wetlands for Water Treatment,* Water Research Centre.

Grant, N., Moodie, M. and Weedon, C. (2000) *Sewage Solutions: answering the call of nature,* CAT Publications, Machynlleth.

Pickford, (1995) *Low-Cost Sanitation,* ITP, London.

Reed, (1995) *Sustainable Sewerage,* ITP, London.

Stauffer, J. (1998) *The Water Crisis,* Earthscan/CAT Publications, London and Machynlleth.

Water Research Centre, *European Design and Operational Guidelines for Reedbed Treatment Systems,* Water Research Centre, UK.

Allotments

Mollinson, B. & Jeeves, A. (1988) *Permaculture: a Designers' Manual,* Tagari Publications, Australia.

Pacey, A. (1978) *Gardening for Better Nutrition,* Oxfam and ITP, London.

Rainwater Catchment and Alternative Water Supplies

Fernando, (1996) *Water Supply,* ITP, London.

Hasse, R. (1994) *Rain Water Reservoirs: above ground structures for roof catchment,* Viewig-GATE Publications, Germany

Kerr, C. (1989) *Community Water Development,* ITP, London.

Pacey, A. and Cullis, A. (1986) *Rainwater Harvesting, the collection of rainfall and runoff in rural areas,* ITP, London.

Postel, S. (1992) *The Last Oasis: Facing Water Scarcity,* Earthscan, London.

Visscher, J.T. and Veenstra, S. (1985) *Slow Sand Filtration, Manual for Caretakers,* International Water and Sanitation Centre, The Hague.

Solar Water Heating and Photovoltaic Generation of Electricity

Boonstra, C. (Ed.) (1997) *Solar Energy in Building Renovation (IEA Solar Heating and Cooling (SHC) Programme),* James and James, London.

The ECD Partnership (1991) *Solar Architecture in Europe: design, performance and evaluation,* Prism Press, Bridport.

Gregory, et al (1997) *Financing Renewable Energy Projects,* ITDG, London.

Keisling, W. (1983) *The Homeowners Handbook of Solar Water Heating Systems,* Rodale Press.

McNelis, B., Derrick, A., and Starr, M. (ed.s) (1988) *Solar-powered Electricity, A survey of photovoltaic power in developing countries,* ITP, London.

Messenger and Ventre (2000) *Photovoltaic Systems Engineering,* CRC Press, UK.

Rozis and Guinebault, (1996) *Solar Heating in Cold Regions,* ITDG, London.

Streib, J. (1992) *Hot Water from the Sun,* Verlag Joseph Margraf, Germany.

Reuse of Buildings

Latham, D. (2000) *Creative Reuse of Buildings,* Dunhead.

Illustration credits

Chapters 1, 2 and 5
All images provided by the author except for the following:
(a) Images on pages 1, 8 and 12 were kindly provided by a Kosovan architectural student studying at the University of East London in 2000 (I have no record of his name).
(b) Lower image on page 9 © Dukagjin Hasimja.
(c) Image on page 120 © Steve Citrone.

Chapter 3
All images © Shirin-Homaan Saadat

Chapter 4
Images © the students noted in each section except as follows:
(a) Image on page 32 © Steve Citrone.
(b) Images on pages 30, 33, 34, 35, 37 top, 39 top, 44 bottom right, 46 bottom, 58 right, 62 bottom, 69, 74 bottom, 92 provided by the author
(c) All 3 images on page 94 © Jean Dumas.

Chapter 6
All images © Tim Booth except the image at top right on page 135 which was provided by the author.

Index